GW00729058

THE ENCYCLOPEDIA OF INVENTIONS

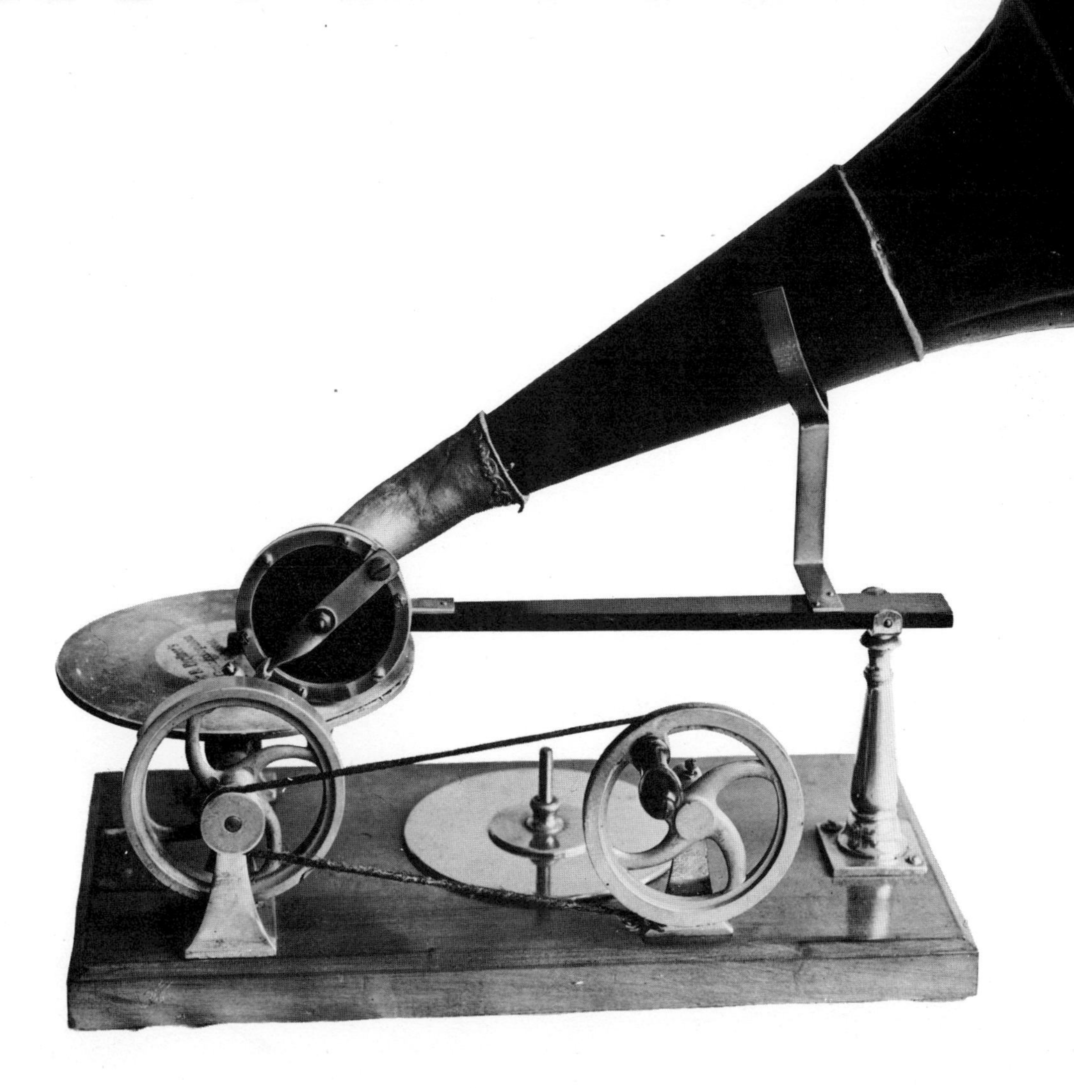

GALAHAD BOOKS • NEW YORK CITY

Edited by Donald Clarke

Library of Congress Catalog Card Number: 76-56963
ISBN 0-88365-390-7
Printed in Great Britain
Published by Arrangement with Marshall Cavendish
Publications Limited

Below: an articulated locomotive, allowing a larger boiler on a single machine. The Beyer-Garratt 4-8-2-2-8-4 was built in Britain for South Africa; it carried 7000 gallons of water, 12 tons of coal and had a mechanical stoking system.

C F KLAPPER

INTRODUCTION

Man is the only animal who knows that he is not immortal. His curiosity about the nature of his earthly existence manifests itself in many ways: as Mark Twain remarked, 'Man is the only animal who blushes. Or needs to.' He is also the only animal who practises religion, who writes novels, and who tirelessly tries to manipulate his environment by using it. Since the first pre-historic man made an axe by shaping a piece of flint, the flow of inventions has increased; technical innovation has perhaps sprung originally from a subconscious desire to live longer by living more easily, but in modern times the reasons for invention have become too many to calculate.

At any rate, it is remarkable how many things which we take for granted had their origin in ancient times. What we usually refer to as the invention of printing was actually the invention of movable type: the Chinese practised block printing for centuries before that. Much of the history of invention is really the history of refinements of earlier invention, and thus it is also the history of economics, philosophy, and of human consciousness itself.

CONTENTS

ANCIENT TIMES 6

Below, left to right: 1) fuel elements in a storage pool in a nuclear reactor: (the blue light is Cerenkov radiation, caused by charged particles passing through the water faster than the speed of light in water); 2) an internal combustion engine, in this case the Rolls-Royce RB211-22 fanjet; 3) Brunel's steamship the 'Great Eastern' in 1857, just before she was launched. She was the largest ship afloat for more than thirty years; 4) the courthouse clock at Ulm, featuring the signs of the Zodiac.

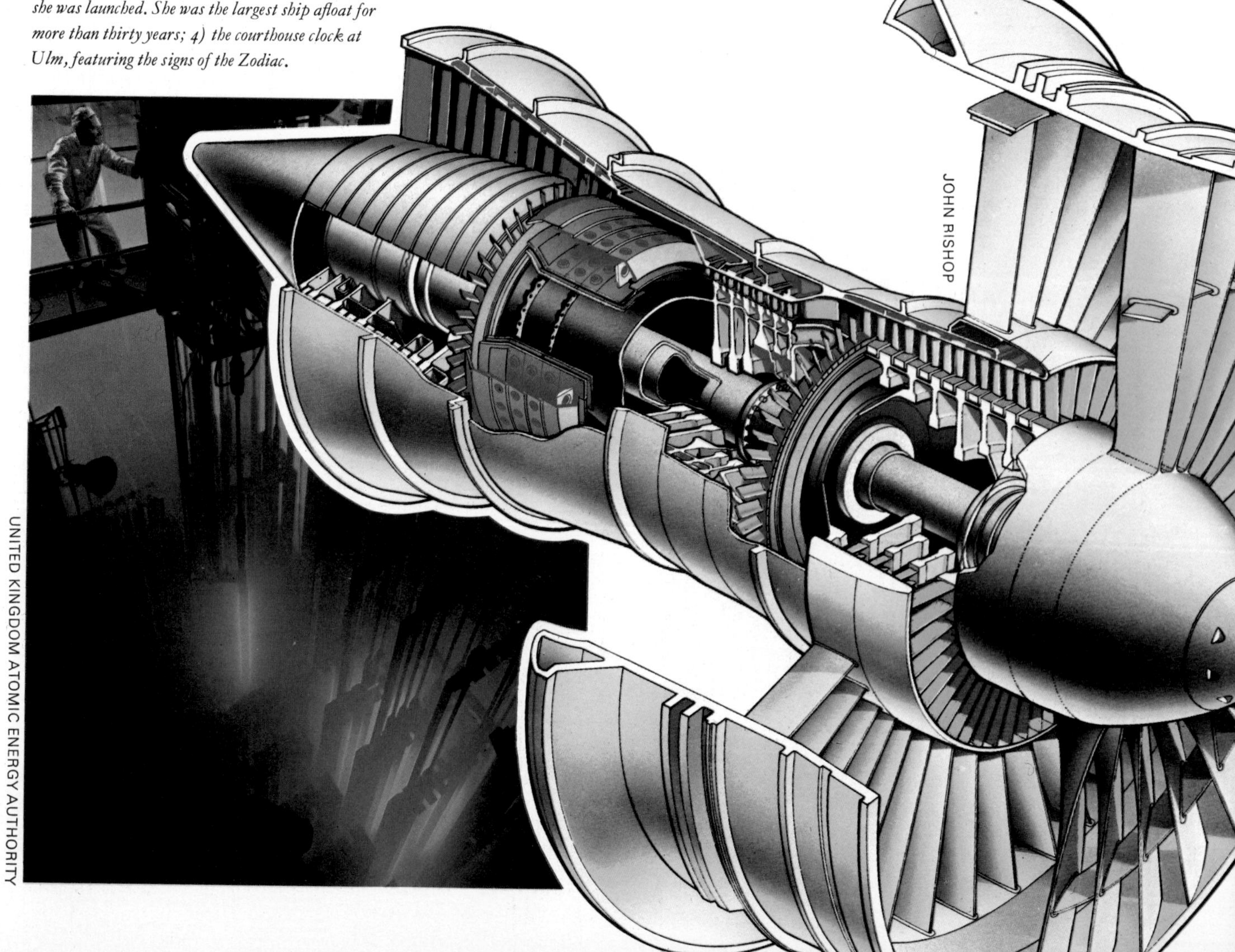

JOHN RISHOP

UNITED KINGDOM ATOMIC ENERGY AUTHORITY

BRUNEL UNIVERSITY LIBRARY

ZEFA

ANCIENT TIMES

When men began to make tools, they created a way of life which was worth recording, and historical time began. The inexorable process of invention and development was already well under way in ancient times. The boat in this picture is made of bundles of reeds: used by Peruvian Indians, it is remarkably similar to boats built by the ancient Egyptians. The attempt to master the environment has been motivated the same way in widely separated parts of the world.

INTERNATIONAL LABOUR OFFICE

The wheel

The invention of the wheel has had a significant effect on peoples throughout the world. Its immediate benefit was to make the transport of goods easier and faster. The improved efficiency in transport, in turn, led to movement out of small villages, where food and vital supplies were within reach, into larger towns, where these goods could be brought in by traders. This change in the way of life from villages to larger towns and cities has been one of the greatest forces in organizing people into civilized communities.

Archaeologists do not know which was invented first, the potter's wheel or the vehicle wheel. Both were originally made of wood and have long since decayed, eliminating any direct evidence. It is obvious, though, when the potter's wheel arrived in a particular culture, because of the fragments of wheel-made pots found during excavations.

The vehicle wheel does not leave traces in this way, and the oldest evidence of its use is a rough sketch of a cart on a clay tablet found in Sumeria (Mesopotamia), dating from about 3500 BC. The oldest known wheel-made pots come from the same area and are about the same age. Part of a clay potter's wheel thought to date from about 250 years later has also been found there.

There is a popular idea that the vehicle wheel was developed from tree trunks used as rollers, but no ancient pictures have ever substantiated this theory. The oldest surviving wheels, found in Mesopotamian tombs of 3000 to 2000 BC, were made of three planks clamped together with cross-

BRITISH MUSEUM – (MICHAEL HOLFORD)

Left: this Sumerian mosaic, dating from about 2100 BC, shows part of a battle and victory scene. The cart shown has wheels made of two planks held together by battens.

Bottom left: a spoked wheel on the Assyrian king Ashurbanipal's chariot. This comes from Nineveh, dating from the 7th century BC.

Below: a Greek relief dating from the 4th century BC showing the race of the Apobatae; the light, spoked wheels were an advantage here.

Bottom: the method of constructing and fixing cartwheels is clearly shown in this 3rd century AD Roman mosaic from Sicily.

C. M. DIXON

BRITISH MUSEUM – (MICHAEL HOLFORD)

C. M. DIXON

struts, and with a natural knot hole in the central plank used as a pivot. The wood around a knot hole is very tough and wear-resistant, which suggests that the wheel turned on a fixed axle, but not enough of the cart has survived to show this.

The first improvement made in the original design was to add a wooden rim to the three planks, which made the wheel wear evenly all around. This rim may have been a single piece steam-bent into a circle, or several segments, *felloes*, joined together. The rim of a Mesopotamian wheel of 2500 BC is studded with copper nails to resist wear; these may have held on a leather tyre. Metal tyres arrived about 500 years later.

At about this time, the spoked wheel was developed, either in Mesopotamia or Turkey. It was first used for war chariots, where its lightness helped manoeuvrability. There may have been a transitional form with a heavy central plank forming the hub and two spokes, and with extra spokes fitted into this crosswise, but only one has ever been found (in Italy) and it is much more recent: about 1000 BC.

By 1500 BC, the Egyptians were building beautifully light four-spoked chariot wheels similar to recent European cartwheels. Wheels with six or eight spokes have also been found in other places.

The design of the wheel became almost completely stabilized around this time, and the only new developments were other uses of the principle, such as the windlass (a machine for hoisting or pulling) and the water mill. It was not until the 16th century AD that any change was made, which was to build *dished* wheels with the spokes arranged in a flat cone. This allowed the tops of the wheels to be angled outwards while the hub still remained directly above the rim. In this way a cart could be built wider but still fit on a narrow track, while the spokes supporting it were still upright and not subjected to extra strain.

Apart from this minor change, things remained much as before until the invention of the *wire wheel*, which was created for the bicycle in the 1870s. The pneumatic tyre was introduced in the late 1880s, although it had been patented 40 years earlier for wooden wheels.

Early cars had wooden spoked wheels, wire wheels or *artillery* wheels, which were one-piece cast-iron copies of wooden wheels. The pressed-steel wheel replaced these in the 1930s; it was lighter, stronger and cheaper to make.

Recently, however, cast wheels have reappeared on the market, and are used on sports cars. They are made of light magnesium or aluminium alloy, and some kinds look similar to the old artillery wheels. In fact, they are very light and strong; their design is based on that of aircraft landing wheels.

SEATTLE ART MUS EUGENE FULLER COLL

RONAN PICTURE LIBRARY

Top: an early T'ang dynasty (7th century) Chinese pottery of a bullock and cart. The wheels appear to have rims, perhaps metal.

Bottom: an early 19th century Yorkshire wagon. The dished wheels allowed for wider loads on the same width of track, and were at a better angle for carrying heavy, swaying loads.

Tools

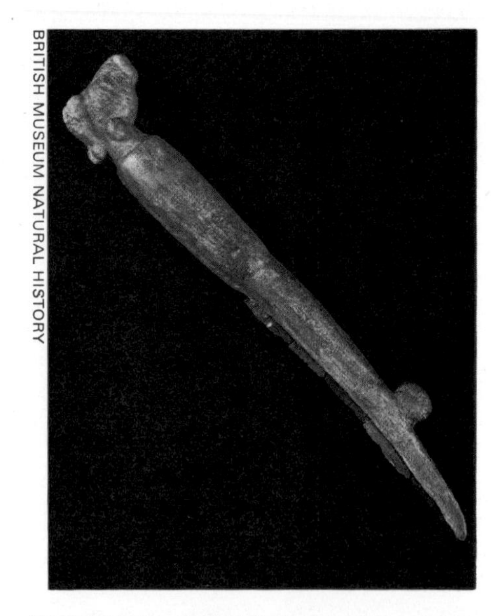

BRITISH MUSEUM NATURAL HISTORY

For thousands of years, early man used simple tools. The hand-axe, for example, was certainly used 300,000 years ago, but it was merely a hand held sharp edge.

It was not until the emergence of modern man, about 20,000 BC, that the first composite tools — devices using more than one type of material — were made. The bow was invented at about this time, and the arrows were tipped with flint points. They were probably held onto the wooden shaft with resin from trees or pitch — the first use of two materials combined.

At this time too it was realised that there were advantages in using handles or shafts on axes, though little evidence remains that this was regularly done.

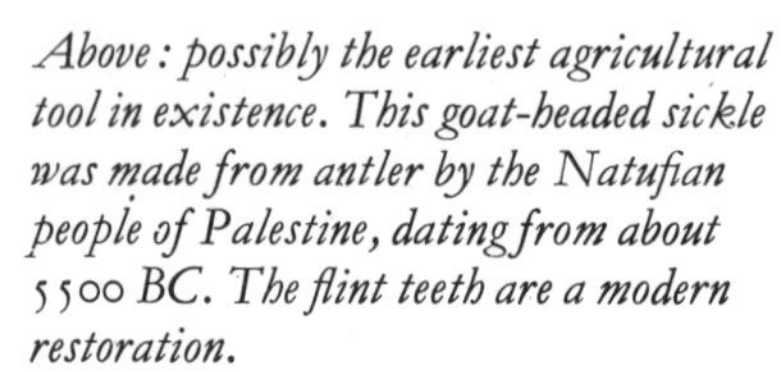

Above: possibly the earliest agricultural tool in existence. This goat-headed sickle was made from antler by the Natufian people of Palestine, dating from about 5500 BC. The flint teeth are a modern restoration.

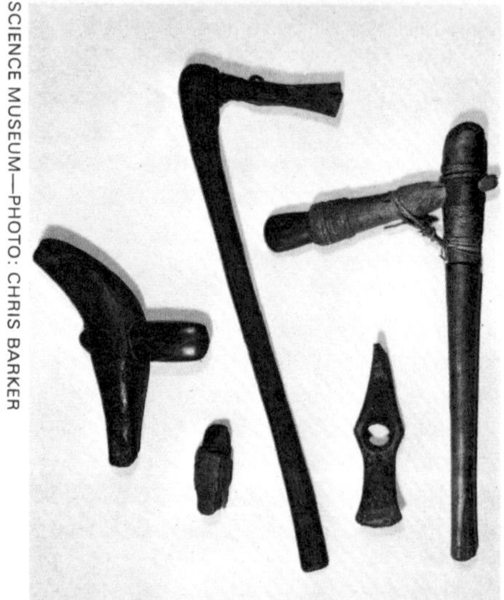

SCIENCE MUSEUM—PHOTO: CHRIS BARKER

Early axes. Left to right: a copy of a Swiss Neolithic stone axe with a wooden handle thick enough to resist splitting; a Stone Age Swiss axe blade with a horn sleeve, also to stop the handle splitting; a Bronze Age axe, an early example of the characteristic "winged" blade shape, with an eyelet for binding the blade to the handle; a copper Moravian combination axe and adze head, dating from about 1800 BC; and a modern primitive Papuan axe with a greenstone blade which can be turned to make an adze.

In the *Mesolithic* or Middle Stone Age (from about 12,000 BC), another type of composite tool, using *microliths*, developed. Microliths are flakes of flint, usually ½ to 2 inches (1 to 5 cm) across, which were stuck into wood, bone or antler to make a variety of effective tools or weapons. Men turned from hunting to agriculture around 8000 BC, and early sickles have been found which use microliths as the cutting edge. Simple saws could also be made in this way. They had the advantage that if one chip became blunted or fell out it could be replaced without having to rework a complete new tool.

When the expanding population began the immense task of clearing the natural forests, the hafted axes and adzes became commonplace.

Early axe and adze blades were made from crudely chipped flint or similar rocks. Flint was by far the most widely used stone because of its tendency to produce flakes with hard, sharp edges. By well directed blows from another stone, a rather rough though quite serviceable blade can quickly be shaped.

If a stone blade is set in a hole bored in a wooden shaft, there is a strong tendency for the handle to split when the tool is used. One solution to this was to set the blade in a sleeve made of antler, which acted as a shock absorber. A ridge on the antler spread the load over a wider area of the handle.

Adze blades are set at right angles to the handle, rather than being in line with it. They are used for dressing and trimming wood, and it is likely that the development of the axe and adze with handles made such innovations as dugout boats possible.

The New Stone Age, or *Neolithic* period, is distinguished by the polishing of the stone implements. Grinding techniques were developed, with sand being used as an abrasive. A high standard of finish was regularly achieved, and the smooth stone cut much more efficiently.

Neolithic men also devised methods for boring holes in stone. They probably used a bow drill — that is, a

LOUVRE, PHOTO: FRANCOIS FOLIOT

A panel at the temple of Ishtar at Mari, Syria, shows a procession of soldiers carrying axes. The temple was founded in about 2500 BC.

bow string twisted round the drill bit. Moving the bow back and forth would spin the bit, which could be a hard wooden rod.

The procedure for making a hole would be to pound out a pilot hole using a hammerstone, and then use an abrasive powder of sand or fragments of flint lubricated with water, with the bow drill. This would tend to wear away a shallow dish, so after a while a new pilot hole would have to be gouged out. The procedure would than be repeated from the other side, with the result that the final hole was narrow in the middle but broad at the edges like an hourglass.

Early pierced blades appear like this, but later tubular drill bits, probably bones, were used to produce a much cleaner cut.

Although this produces a long-lasting axe, not very many Neolithic pierced blades have been unearthed. They were generally reserved for use as battle-axes.

Stone hammers made in this way have also been found, though even well into the Bronze Age workmen still regularly used hammerstones without a haft.

Bronze was first used as early as 3000 BC, and by 1500 BC its use had spread widely. Its high cost and rarity meant that it tended to be used for ornaments and weapons rather than for workshop tools, and stone continued to be used even into the Iron Age.

Curiously enough, once metal tools had been seen by craftsmen, they set about making their polished stone counterparts as similar in appearance as possible, right down to the seams of the castings.

PHOTO WILLIAM MACQUITTY

VICTORIA & ALBERT MUSEUM

Above: this illustration of boat building is in the tomb of Ti, an official who lived in Egypt about 2400 BC. Composite adzes of simple design are being used.

Left: the Bronze Age came suddenly in China, about 1600 BC. Before this date only materials such as stone and wood were used; after, elaborately cast bronze weapons like this axe head from the late Shang or early Chou, around 1100 BC, are found. The technique was probably learned from the west, but modifications in casting were made which improved the quality.

BRITISH MUSEUM

ZEFA

Above: ancient Mexican sacrificial knife of chalcedony. It was made by the Mixtec culture in the 14th century AD.

Right: digging sticks, like these being used near Cuzco, Peru, were the forerunner of the spade. Holed stones could be added to the shaft to give extra weight for digging in difficult soils.

Numerals

The ancient Egyptians were among the earliest people to evolve a kind of decimal system for recording numbers. Units were written as a single vertical stroke, so that four units were recorded as four strokes. A new symbol, like an inverted letter U, was used to denote ten units, while other symbols were employed to record hundreds, thousands, ten thousands, and so on. Thus, to write a number such as ninety-eight, the ancient Egyptians had to set down eight strokes and nine symbols for the number ten.

The Romans slightly improved upon this system in two ways. First, they introduced additional symbols to denote five, fifty and five hundred; and secondly, by placing a symbol of a lower value in front or behind one of higher value they showed whether it was to be added or subtracted to obtain the required number. Thus L denoted fifty, XL denoted forty (fifty less ten) and LX sixty (fifty plus ten). At much the same time the Greeks produced a numerical system in which the letters of their alphabet were written with a stroke above to indicate the numbers one to ten, and thereafter twenty, thirty and so on.

The great disadvantage of all these systems was that it was impossible to make any but the simplest arithmetic calculations. Single additions or subtractions presented no difficulty, but to multiply two large numbers was quite a different matter. The practical solution was to use a sand-table, called an abacus, named after the Greek word for sand, *abax*, and not to be confused with the later bead-frame of the same name first used by the Chinese. A grid was first drawn in the sand with vertical, horizontal and oblique lines. The numbers to be multiplied were then set out as marks in the sand along the top and down the side of the grid (fig 1). Each column was then multiplied, place by place, against the other, units being recorded below the oblique line and tens above it in each square of the grid (fig 2). The oblique columns were then added and the numbers set down (fig 3) along the side and bottom of the grid. The answer obtained still had to be transliterated into Roman or Greek numerals. This unwieldy procedure made complex calculations very difficult and they were often the work of experts.

By 500 AD astronomy had reached a point where complex mathematics were essential, and this was especially true of the work being done in the cities of the Punjab, at that time a melting-pot of Hindu, Buddhist and Greek cultures. Shortly after 500 AD, according to tradition, the astronomer Aryabhata decided to replace the cumbersome marks made on the abacus by a group of symbols representing the numbers one to nine. He realized, furthermore, that the actual numerical value given to the symbol could depend upon its position. Thus the symbol for three could mean three units in the first

MICHAEL HOLFORD LIBRARY

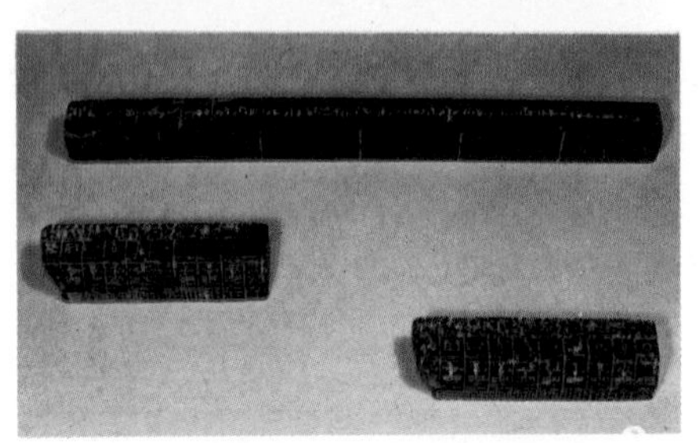

DIANA WYLLIE LTD

Above left: in this cuneiform tablet from Mesopotamia, numbers are represented by rows of nicks in the clay smaller than the ordinary characters.

Left: fragments of Egyptian measuring rods of the 16th century BC. Number symbols are clearly visible along the bottom of the two shorter pieces.

BERRY FALLON DESIGN

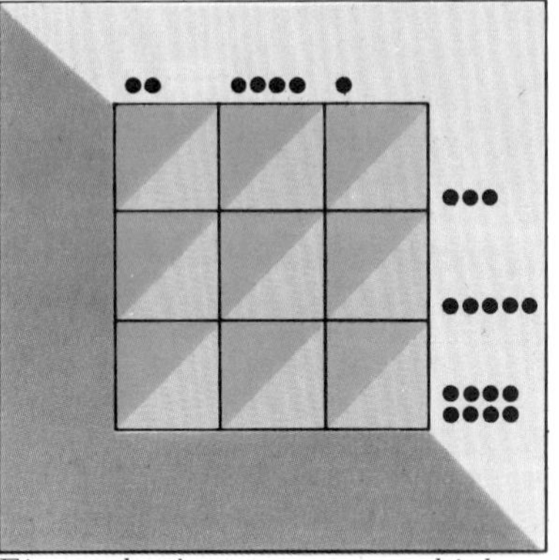

Fig. 1: the abacus set out to multiply 241 by 358 (see text).

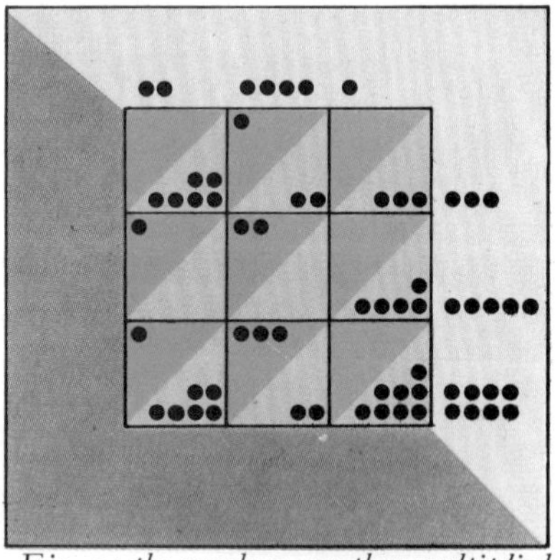

Fig. 2: the numbers are then multiplied together square by square.

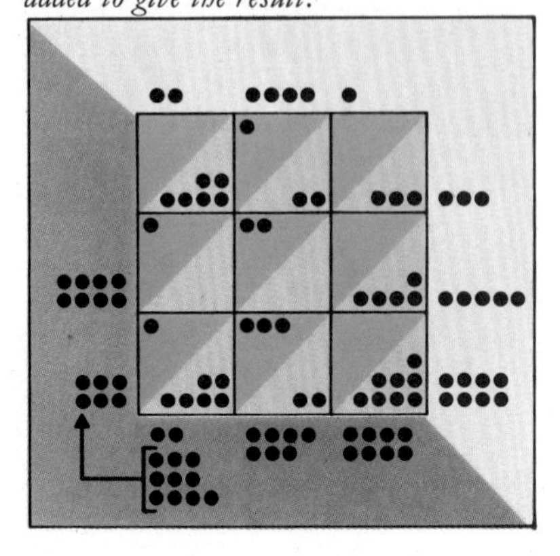

Fig. 3: finally, the oblique columns are added to give the result.

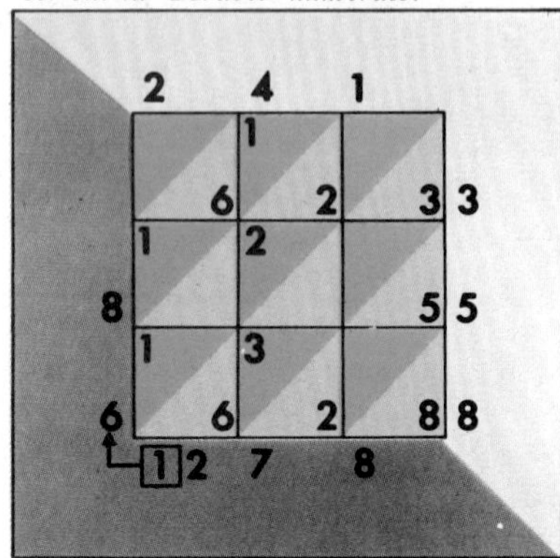

Fig. 4: the same multiplication sum set out in 'Arabic' numerals.

column, thirty in the second column, three hundred in the third column, and so on. In other words, he introduced the idea of place notation that we use today, an idea already present in embryo in the practical use of the abacus. One further improvement had to be made. In using the abacus it was perfectly practical to leave a blank space in a column; and in setting the figures down as Roman numerals there was no need for the figure zero. Hence, CV is read as 105, and there is no need to record that no tens are present in the number. With place notation, however, a symbol was required to show the place was void, and the Indian scholars introduced a sign for zero, the *cipher*.

When the Arabs conquered the Punjab shortly after 700 AD, they were impressed by the advanced state of Indian mathematics and, in 771 AD, Indian scholars were brought to Baghdad to teach the new system. Their symbols were adopted by the Arabs, who referred to them as *Hindi* (Hindu) or *huruf alghubar* (sand letters). Soon not only Arab scholars but also traders and businessmen were using *Hindi* symbols. Europeans were slow to learn from the Arabs. Although a few European scholars had adopted the system before 1200 AD, its general use did not gain popularity until after this date when the *Hindi* figures, known in Europe as *Arabic* figures, were introduced by Leonardo Fibonacci of Pisa. A century later, when the first true clocks began to be built, Arabic numbers were still not widely understood, and Roman numerals were used.

PHOTO: PETER CLAYTON

Above: a Roman inscription including numerals. Note the form XVIIII for 19, which was later written XIX.
Below: how numerals developed, from Egyptian hieratic, a handwritten development of hieroglyphic, to today. The first two rows are written from left to right, in defiance of actual practice, to make them easier to follow.

EGYPTIAN HIERATIC	[illegible]	[illegible]	[illegible]	[illegible]	[illegible]	[illegible]	[illegible]	[illegible]	[illegible]	[illegible]	[illegible]	[illegible]	[illegible]	[illegible]	[illegible]	[illegible]
HEBREW	א	ב	ג	ד	ה	ו	ז	ח	ט	י	יא	יט	כ	כא	ק	קכו
GREEK	A	B	Γ	Δ	E	F	Z	H	Θ	I	IA	IϘ	K	KA	P	PKF
ROMAN	I	II	III	IV	V	VI	VII	VIII	IX	X	XI	XIX	XX	XXI	C	CXXVI
HINDI	१	२	३	४	५	६	७	८	९	१०	११	१९	२०	२१	१००	१२६
MODERN ARABIC	١	٢	٣	٤	٥	٦	٧	٨	٩	١٠	١١	١٩	٢٠	٢١	١٠٠	١٢٦
MEDIAEVAL EUROPEAN	1	2	3	4	5	6	7	8	9	10	11	19	20	21	100	126
MODERN	1	2	3	4	5	6	7	8	9	10	11	19	20	21	100	126

BERRY FALLON DESIGN

Writing

C M DIXON

Above: one of the earliest existing written records, a stage away from Mesopotamian picture writing. This clay tablet is an account of fields and crops, and dates from about 2800 BC. From this developed the cuneiform script, as shown below: a sample from Persepolis, about 500 BC.

The history of writing began in Mesopotamia in about 3500 BC, when the palace administrators found it essential to keep records of property, dues and taxes. The cheapest and most easily obtainable material on which to make these records was clay, which could be rolled out into thin tablets, drawn on while still damp, and then dried in the sun. The first scripts were word-signs: every object had its own symbol, normally a simplified picture of the object itself. Originally the word-signs were drawn with a pointed stick. In a very short time a large number of symbols had been designed in order to make accurate recording possible. There was, for example, no single word-sign for sheep but different symbols to describe rams, ewes, lambs, and so on. The system rapidly became unmanageable as more and more symbols were added.

So to simplify the Mesopotamian script two important steps were taken. First, the original method of drawing word-signs was abandoned, and the symbols were created by jabbing the surface of the tablet with a piece of reed that was naturally triangular in section. In this way word-signs were built up from a number of wedge-shaped impressions in the clay to give what is known today as *cuneiform writing* (from the Latic *cuneus*, 'wedge'). In the second development, the number of symbols was reduced by the use of 'determinative' signs which had the value of adjectives. As a result, only a single symbol was needed for 'sheep' to which could be added determinative signs to show whether it was male, female or a lamb.

The well known Egyptian *hiero-*

Below: Egyptian hieroglyphs, dating from about 1450 BC, recording the restoration of part of a temple.

ZEFA

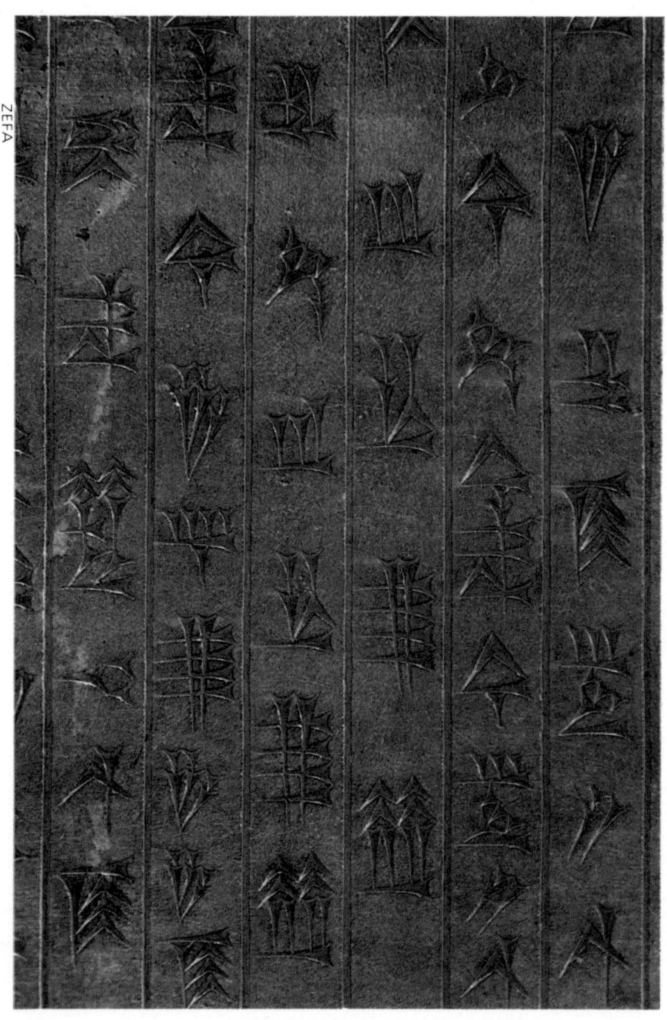

SCALA

glyphs were also word-signs used in much the same way as those of Mesopotamia, although they appeared later. At first they were used to record important events and so were designed to be carved on stone surfaces. For this reason they appear neater than the Mesopotamian symbols, and are, indeed, more detailed. As in Mesopotamia the Egyptians took to using determinative signs to show what class of object was intended.

Another important step taken in Egypt and Mesopotamia was to use the word-signs as syllables, a single word being built up from two or more symbols, as though in modern English, for example, the word 'wagon' was written as two signs, one meaning 'wag' and the other 'on'. In neither country, however, was this development carried to its logical conclusion—a writing based entirely on signs each with its own sound—a phonetic script.

With the invention of Egyptian papyrus, an early form of paper made from reeds, in about 3000 BC, the old form of hieroglyphs had to be changed to give a more rapid system of writing. Each symbol was reduced to a few lines to give what is known as *hieratic* script. This was the beginning of our modern system of writing, for the old cuneiform was so totally adapted to recording on clay tablets it could not be altered to write with brush or nib on paper.

By 1000 BC the Phoenicians, the great trading people of the eastern Mediterranean, had borrowed elements from both Egypt and Mesopotamia, and had created their own script in which each consonant was given a separate symbol, but there were no vowels. In modern English it is as though the words shop, ship and sheep were written as two symbols, one pronounced 'sh' and the other 'p'. Only the context could reveal which word was intended. The Phoenician script had the advantage that, like hieratic, it could be written rapidly.

The early Greek colonists took over the Phoenician script and by 600 BC improved its major shortcomings by adding a number of signs to express the vowels, and created what was to become the alphabet used in common by all European countries.

Of the other major scripts in use today Arabic, like the Phoenician from which it was developed, originally had no vowel signs, although today they are sometimes added. The Chinese system of writing, on the other hand, followed a somewhat similar development to that of Egyptian hieroglyphs. The symbols made with brush and ink are in fact word-signs simplified from earlier more pictorial forms. It was the invention of paper around 100 AD that brought about this change, but largely because words were pronounced so differently in local dialects, the symbols never became phonetic but retained their meaning as word-signs.

MICHAEL HOLFORD

C M DIXON

Top: This is hieratic script, a later written form of hieroglyphs. This sample probably dates from about 1250 *BC.*

Bottom: early Greek writing used many letters familiar to us today. This form of writing, however, known as boustrophedon, marks a transition in the direction of writing: the lines read alternately left to right and right to left. About 600 *BC.*

Paper

The earliest writing materials to be used in the Old World (by the Mesopotamians more than 5000 years ago) were thin tablets of clay. Although cheap and easy to produce they had two major disadvantages: they were difficult to store, and could only successfully be used by impressing the writing tool into the surface when the clay was soft. The papyrus reed, which grew profusely in the delta of the Nile, provided a far superior writing material. The tall papyrus reed is triangular in section, and it is possible to peel off long, narrow strips of white fibrous pith which, when fresh, contains a sticky gum-like substance. Strips of papyrus pith had been used in the Egyptian New Stone Age for making mats and baskets, but about 2500 BC the discovery was made that they could provide an ideal writing material. It is from the word papyrus, through Greek and French that the word 'paper' is derived.

In the manufacture of papyrus, strips were laid on a flat surface, edge to edge, and across these was placed a second layer of strips at right angles to the first. The two layers were then beaten with a heavy mallet to flatten them, and allowed to dry in the sun. Sheets of papyrus formed in this way were then joined edge to edge, and could be rolled up for storage.

Almost as ancient as the making of papyrus was the manufacture of parchment from the skins of new-born lambs or calves. Early parchments differ little from the finer leathers produced during these ages, but during the first thousand years BC a technique for making a fine, white writing material was evolved. The skins were first soaked in a bath of lime, and the hair was then scraped away. They were subsequently returned to a fresh vat of lime water, after which they were stretched on a frame to dry. During the drying process the skins were carefully thinned down by paring away the surfaces with a half-moon knife. Finally, when the skins were dry, the surfaces were smoothed with fine pumice powder such as a dentist might use today to clean teeth. The parchment was then cut into convenient rectangular shapes, and these, as in the case of papyrus sheets, were joined edge to edge to form scrolls.

In the first century AD it was realized that parchment could be stored in a far more convenient manner than as rolls. Each rectangular sheet could be folded once, twice of three times, and the edges cut to provide a folio, quarto or octavo.

Below left: because of the dry Egyptian climate, some very early papyrus sheets have survived. This one, dating from before 1200 BC, shows a stylized papyrus plant.

Below right: a parchment scroll in both Greek and Latin, of Byzantine origin.

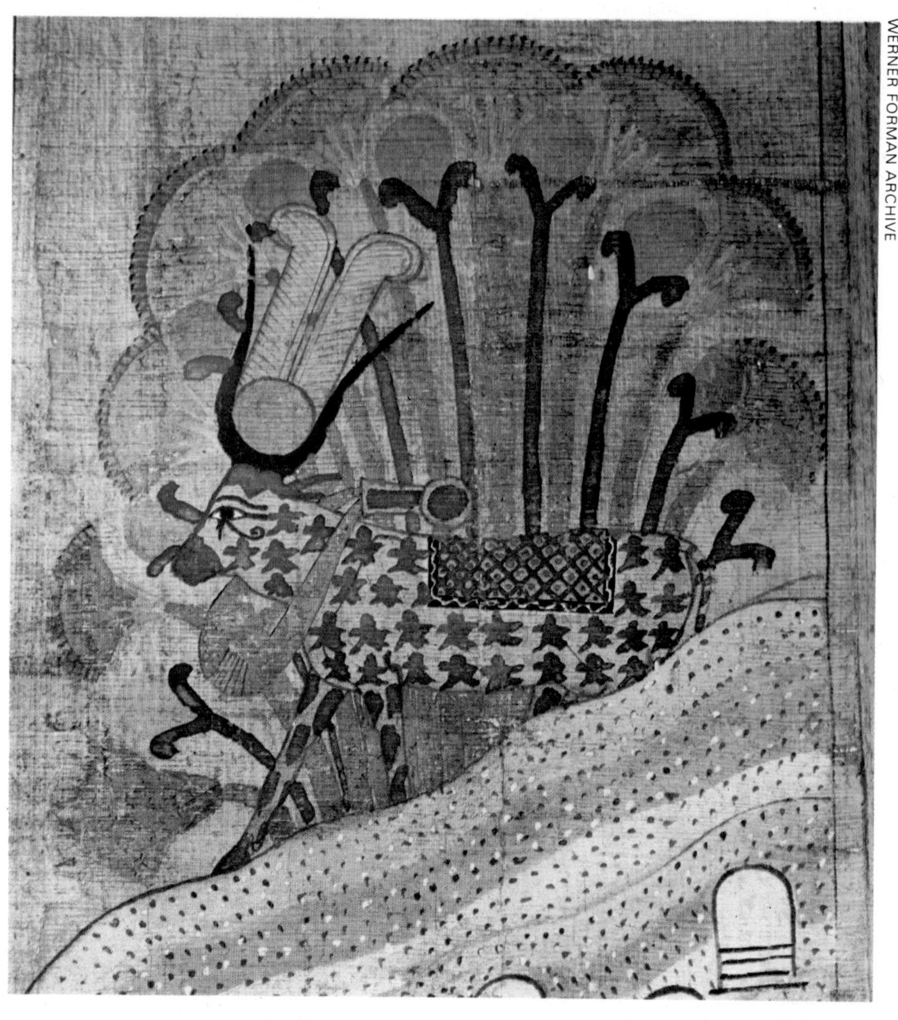

WERNER FORMAN ARCHIVE

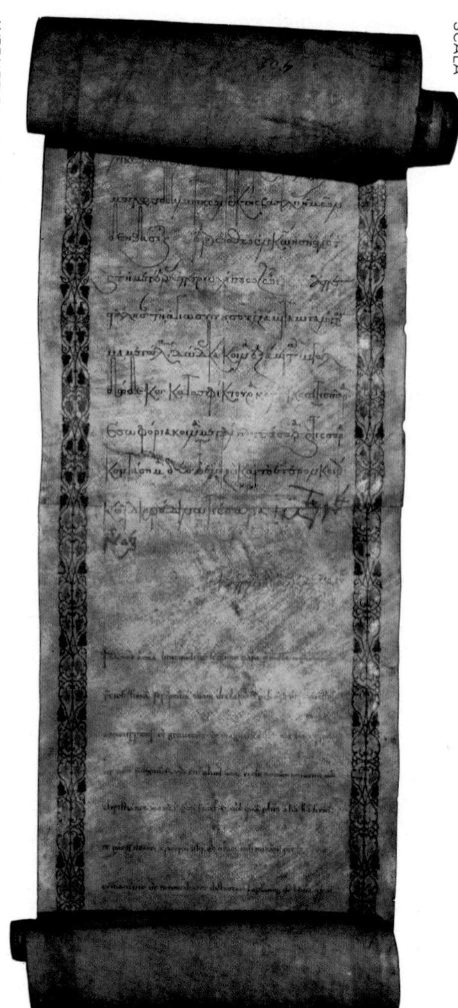

SCALA

Such sheets of parchment could then be stored by being bound between thin, flat, sheets of wooden board. This *volumen,* as it was called, provided the origins of the book as we know it today.

In the first century AD the Chinese also began to manufacture paper. Initially it was made from the fibrous *phloem* that underlies the bark of the mulberry tree, and there seems little doubt that the idea was evolved from the manufacture of felt, widely practised by China's nomadic neighbours in Central Asia. Just as felt was made by compacting animal fibres into a sheet, so paper was made by compacting plant fibres. In the process as it developed ultimately the mulberry fibres were mixed with water to form a mash which was then shaken into an even layer covering the surface of a fine sieve. After it had drained for a while, the sheet so formed was carefully picked up by the corners, smoothed out against a wall in the full sun and left to dry.

By the middle of the eight century AD the Chinese, who had been spreading their empire westward into Central Asia, came into conflict with the forces of Islam which had been spreading eastward. The Mohammedans took a large number of Chinese prisoners, and from them learnt the art of papermaking, establishing factories in Samarkand. The technique was to spread rapidly in the Islamic countries, first to Damascus, then to Egypt, North Africa and Spain. By this time the original mulberry fibres had been replaced by many substitutes, the most important of which were esparto grass and linen, usually in the form of rags.

In the thirteenth century paper began to be made in several European countries. Until then, in China and these countries of Islam, the process had been carried out entirely by hand, but in Europe, largely because the only readily available raw material was linen waste, water mills were adapted to reduce the rags to pulp. Thus, while the paper was still laid by hand, by the late Middle Ages some part of the papermaking industry had already been industrialized.

SONIA HALLIDAY

ZEFA

Top: Chinese papermaking in the early 19th century. The pulp is heated and then spread on a porous table to set. The angled slabs are heated for the final drying.

Right: a parchment book of the 8th century from the monastery of St John on the Greek island of Patmos.

The bow

The bow is not quite the world's oldest machine, but it is the oldest machine with more than one part, and the oldest energy-storing device. It had an enormous influence on the development of early man, turning him into a really efficient hunter. For the first time, he could bring down an animal further away than he could throw a spear. This meant that his inability to run as fast as the swifter animals ceased to matter.

A bow works by storing the full strength of a man's arm, fed into it as it is bent, and then releasing it in a single explosive moment. The energy transferred to the arrow in this way moves it far faster than it could be thrown by the arm alone. Not only the speed of the arrow is increased, but also its range. A good modern longbow will send an arrow 400-500 yards (or metres).

It is hard to say exactly how the idea of the bow came about (perhaps from observing the springiness of saplings) but its immediate ancestor was the *spear thrower*. This extremely primitive device is still used by Eskimos (for throwing harpoons) and by Australian aborigines (who never invented the bow, since they had the boomerang).

The spear thrower is a wooden stick about 2 feet (60 cm) long with an inward-facing cup hollowed out of one end, so that it is like a ladle with a thick shaft and a small bowl. The user points his arm backwards, holding the thrower by its plain end to form an extension of his arm. The butt of the spear shaft is fitted into the cup.

The spear is thrown by swinging the arm and thrower forward together. The effect of the thrower is to make the user's arm 2 ft longer, so that more velocity can be given to the spear.

The bow was a natural development from the spear thrower, not because it resembles it as a machine, but because they are both methods of artificially increasing the speed of a projectile.

The date of the invention of the bow is unknown but the oldest representation of it is in a north African cave painting that dates from 30,000-15,000 BC (towards the end of the Old Stone Age). The oldest surviving bow dates from the Middle Stone Age (10,000-3,000 BC) and was found preserved in a peat bog at Holmegaard, Denmark.

Like all early bows, this is one of the type known as a *self bow,* that is to say it is made of a single piece of material (in this case, elm wood). Later bows are often *compound* or *built* bows, made of several pieces joined together, or *composite,* made of several different materials arranged so that their particular characteristics give the bow extra strength and spring. Eskimo composite

POPPERFOTO

BRITISH MUSEUM-PHOTO: MICHAEL HOLFORD

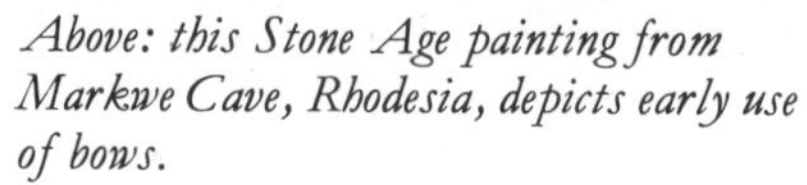

Above: this Stone Age painting from Markwe Cave, Rhodesia, depicts early use of bows.

Left: Pharaoh Rameses II (1290-1223BC) shown using a wooden self bow.

WILLIAM McQUITTY

Right: this 4th century plate shows a Parthian using a recurved bow while retreating on horseback — a tactic which gave rise to the phrase 'a Parthian shot'.

BRITISH MUSEUM-PHOTO: MICHAEL HOLFORD

A sixteenth century Florida Indian with an unusually long bow and flint headed arrow, from a drawing by John White. Bows were not generally made longer than their users.

bows, for example, are made of reindeer antler reinforced with plaited animal sinews along the front edge. Modern ones often use metal and fibreglass.

The length of the Holmegaard bow is 57 inches (1.45 m), about the same length as a modern bow. But bow lengths vary widely, from less than 3 ft (91.4 cm) as used in Africa to over 8 ft (2.44 m) in Japan. The classic English longbow, as used at the battles of Agincourt and Crecy, was a self bow of yew wood the same length as the user.

This bow had a string of linen or hemp, but any tough, non-stretching material can be used, including silk, rawhide and even a bamboo strip. The string must not stretch because all the elasticity is provided by the bow stave itself. It is normally attached by being looped round a notch or shoulder in the ends of the bow, but some primitive bows have holes in the end through which the string is threaded.

Various improvements have been made on the design of the bow at various times. One of these was the *recurved* bow made of wood, horn and sinew and intended to be used from horseback. Its name comes from the fact that the ends of the bow stave curve forwards, against the curve of the rest of the bow. This double curve gave it great elasticity and a corresponding range and power. Modern composite bows are also recurved.

Another development is the crossbow, which was probably invented in China, and was much used on the mainland of Europe during the Middle Ages. It had a short but powerful bow set horizontally on a stock like that of a shotgun, and often needed such strength to draw that it had to be wound up by a windlass or similar mechanism. The string was released by a trigger.

The conventional longbow shot arrows made of wood, cane or reed, often with a head of metal or flint. They normally had *flights* of feathers or other light material at the tail end to steady them. The crossbow shot shorter, heavier *bolts*. Some types shot small stones or bullets.

The bow lost its importance as a weapon in Europe with the introduction of firearms towards the end of the Middle Ages. The longbow is now used only for sport and target shooting, except among primitive peoples. Modern sporting longbows may be reproductions of the traditional English yew self bow or composite recurved bows made of various materials. The crossbow survives as a target weapon; modern ones have alloy bow staves.

BY PERMISSION OF THE TRUSTEES OF THE NATIONAL GALLERY

'The Martyrdom of St Sebastian' was painted by Antonio Pollaiuolo, who lived from 1430 to 1498. The bows shown here are typical of those used in Italy at that time: the man at the back is shooting with a recurved bow, while others are loading and using various crossbows.

BY PERMISSION OF THE TRUSTEES OF THE WALLACE COLLECTION

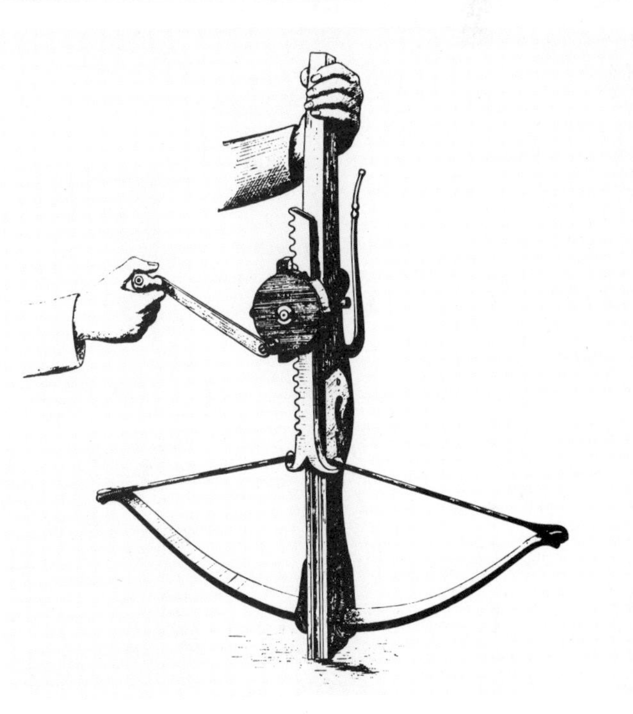

PHOTO: DAVID HOFFMAN

Top: a German fifteenth century crossbow, with the bow itself made of steel covered with parchment, and the stock of wood with horn veneer. Bottom: the heavy pulls required to draw crossbows were often given with a cranequin, a rack and pinion device.

Printing

The idea of printing appears to have been developed in India in the few centuries immediately before the birth of Christ. India had long been renowned for the quality of its cotton cloth and the skill of its dyers. The Greek writer Strabo, who died in 20 AD, described the printed textiles from India. Apparently the pattern to be printed was cut in relief on the flat surface of a wooden block, which was then covered with dyestuff and pressed down upon the stretched cloth. Such blocks were quite small and used to create simple repeat patterns.

By the sixth century AD the art of textile printing had reached Egypt. Examples of such printing and the wooden blocks used in the process have been found in the ruins of the ancient city of Panopolis, modern Akhmim. Another example of printed Egyptian cotton has been found in the tomb of St Caesarius of Arles, who died about 543 AD.

The art of textile printing did not only spread westwards, however, for equally early examples are known from Japan, Persia and Central Asia. Undoubtedly knowledge of this work was brought to Central Asia by Buddhist missionary monks who, from the second century AD onwards, became increasingly active in this area. It was here that they came across elements of Chinese culture that were to prove an enormous stimulus to the future development of printing.

By 500 AD the manufacture of paper had become a well established practice in China and the secret of its production had clearly spread to the region of Sinkiang. Furthermore, the introduction of paper had led to the development of a new writing medium called Chinese ink. Traditionally this was invented by Wei Tang in about 500 AD. The material was a mixture of lampblack and a water soluble gum or size. It was made up as a solid block from which a liquid ink could be made by the addition of water as required.

In a manner of speaking, however, the Chinese themselves had already evolved a form of printing. It had become the accepted custom to use a personal seal in order to validate documents and letters, much as we use signatures today. At first the characters were cut into the surface of the seal. The seal was covered

MICHAEL HOLFORD LIBRARY

Above: two Chinese imperial jade seals. Characters are cut into the surface, which is covered in greasy red 'ink' and pressed on paper to make an impression.

Below: book printing in the 16th century was a slow process: each letter was set individually, the page was printed and then hung up to dry the wet ink.

BULLOZ

with a mixture of the red pigment, cinnabar, and grease or oil and pressed on to the paper. This gave an impression of white characters against a red background. By 400 AD a significant change had taken place, for by then it was more common to cut the characters in relief, so that when printed they appeared red against a blank background.

Chinese ink could not be used with seals of metal or precious stone since, being water-miscible, it would not wet the surface properly. It was probably the Buddhist missionaries who first discovered that Chinese ink could be used with a wooden printing block, with which they were already familiar. Hence some of the earliest pieces of printing known come from Central Asia. These take the form of slips of paper on which were printed an image of the Buddha, a simple form of charm, and were probably made in about 600 AD. In the next century Chinese historians record the use of printing blocks in both India and China, while in 767 AD the Empress of Japan, a devout Buddhist, ordered a million charms to be printed.

For small items of printing the wood block was perfectly satisfactory. The picture or text was cut in reverse on the face of the block, which was then inked. The paper was placed on the block and smoothed down with a hemispherical pad. This is a reversal of the order in which printing is done today. By the tenth century the Chinese were printing not only books, but also calendars, playing cards and even paper money.

The earliest surviving printed 'book' (actually a scroll) is the Diamond Sutra, produced about 868 AD. It is approximately a foot deep and when unrolled, about sixteen feet long (30×490 cm). Each 'page', or section of text, was printed from a separate block, which had to be cut by hand. Even so the Buddhist canon, the Tripitaka, was printed as 5048 scrolls in about 980 AD, while the Taoist canon of 4565 scrolls was completed circa 1019 AD.

Block printing did not survive in Egypt, and the idea had to be reintroduced into the countries of Islam and Europe. By 1294 AD paper money was being printed in Tabriz in Persia, while the earliest known printed version of the Koran, from near Al Fayyum in Egypt, is of much the same date. By 1400 AD playing cards and textiles were both being printed in Europe, and the oldest surviving picture print known is one of St Christopher made in 1423 AD.

PHOTRI

Above: Japanese printmakers used the wood block method of printing. The picture or text was cut in reverse on the face of the block, which was then inked.

Below: simple repeat patterns are typical of Indian printed textiles and are cut in relief on a small block, covered in dye and pressed on to stretched cloth.

BURY PEERLESS

Mapmaking

Sketch maps and unscaled plans of small regions, estates and cities are known from the civilizations of the Near East from as early as 2500 BC. It was not possible to create maps, as we understand them today, until a system had been devised whereby the spatial relationship between one point and another could be recorded, and that had to depend upon astronomical observations. Not until the centuries immediately after the birth of Christ was astronomy sufficiently advanced to allow the measurements of the two vital co-ordinates, longitude and latitude, by which a position could be fixed.

A second difficulty that lay in the way of accurate cartography was the then accepted method of measuring long distances. Travellers would refer to a city as being so many days' journey from another city, and since the nature of the road between the two points decided the time taken to travel it, the distance covered in a day's march might vary considerably.

The father of cartography is usually considered to be the Greek geographer Ptolemy, whose maps were compiled in Alexandria in about 150 AD. The success of his work, although depending upon the ideas of previous geographers, lay in the systematic introduction of the two co-ordinates as a means of plotting a position. For measurements of longitude he took Alexandria as the meridian and measured the distance that a city lay east or west of that line, the distance being calculated on the difference of time between sunrise in that city and at Alexandria. Equally, measurements of latitude were made by comparing the height of the sun at midday in a particular city with the height of the sun at Alexandria. This had to be carried out at an equinox, in March and September, when day and night are equal. That Ptolemy's maps were inaccurate is hardly surprising: he had no timepieces, other than simple water clocks, to aid him in his calculations; often he had to rely upon the observations made by others, which might be very wide of the mark; and he had no magnetic compass with which to cross-check locations. Even so, there is some evidence to suppose that Ptolemy used a system of triangulation as a means of checking positions. Thus, if he was in doubt about the position of a

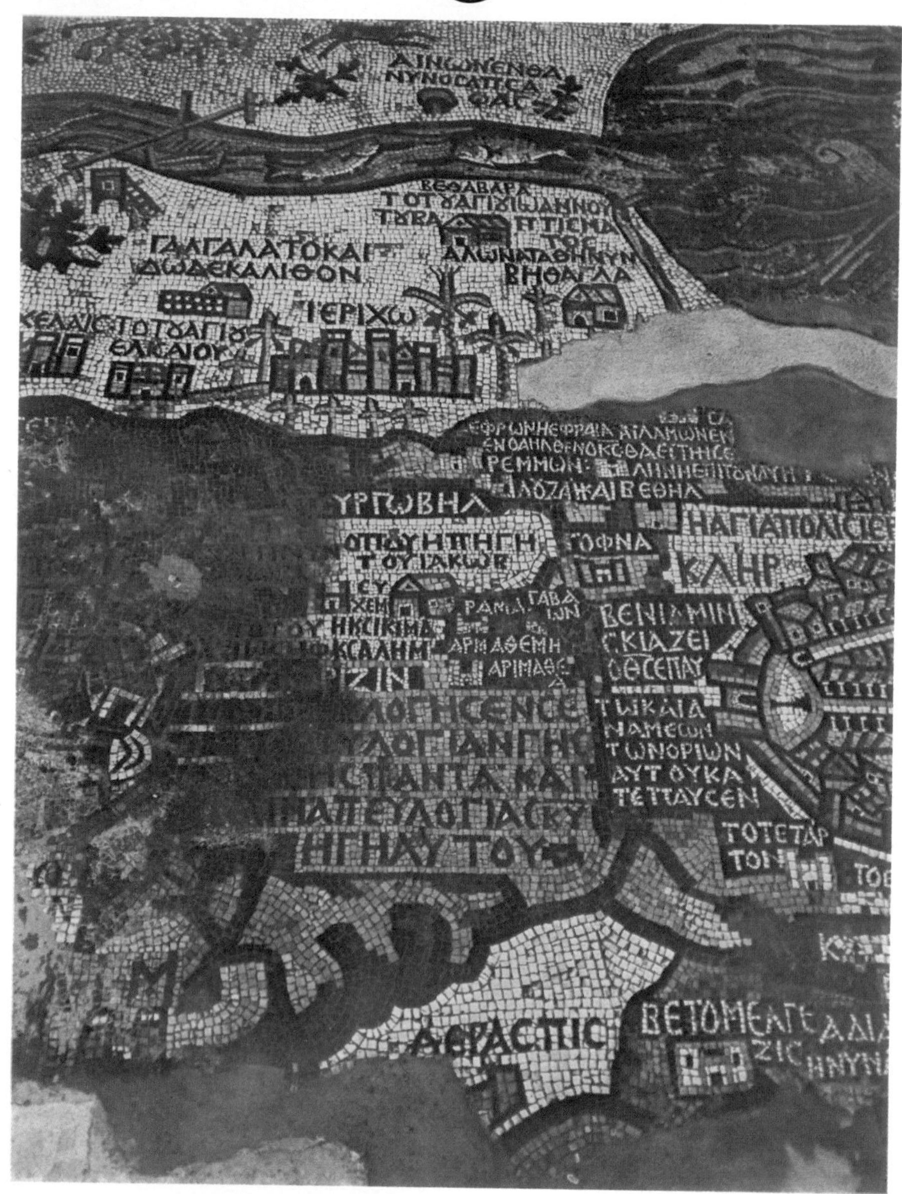

CLAUS HANSMANN

BRITISH MUSEUM/PHOTO: C M DIXON

Above: a mosaic map of Palestine and Jordan from the floor of the Greek Orthodox church of Madaba in Jordan. The map dates from the 6th or 7th century AD. *The town of Jericho, marked Ιεριχω in Greek, is in the centre; Jerusalem is at the lower right corner.*

Left: an early example of a map inscribed on a stone tablet dating from the 6th century BC. *The tablet was discovered at Sippar, an ancient city of Babylonia, and was drawn to illustrate the campaigns of Sargon of Agade (c. 2300), an ancient ruler of the Near East. The map shows the oceans and has Babylon, on the river Euphrates, at its centre.*

city as given by its reported co-ordinates, he could test its position by plotting the known distances between it and other cities in whose location he had more faith.

Despite their defects Ptolemy's maps remained the basis upon which most Arab and Western cartographers depended during the Middle Ages. His work was translated into Arabic and considerably improved upon by Islamic geographers from the tenth century onwards. Thus, al-Istakhi, a scholar in Baghdad, produced a revised version of Ptolemy's maps in colour in about 950 AD; while two centuries later al-Idrisi, working at Palermo, Sicily, created a disc-shaped map of the world.

The countries of the Near East were not alone, however, in showing an interest in cartography. At the same time that Greek scholars were laying the foundations upon which Ptolemy was to build, Chinese geographers, too, had made similar advances. In 271 AD P'ei Hsin, the minister of works to the Emperor, produced a map of China in eighteen sections. Only the introduction to this work survives today, but in it P'ei Hsin, whom the Chinese equally look upon as the father of cartography, laid down a set of rules for map makers. The map was set out on a grid system, positions being located accurately within the lines of the grid. Thus, to determine the longitude of a city, a post was set up to stand eight feet (2.5 m) above the ground, and the length of its shadow measured at noon at an equinox. If the shadow length recorded in this way showed a difference of one inch (2.5 cm) between two places they were reckoned to lie at a distance of 330 miles (100 km) north and south of one another. P'ei Hsin insisted that roads should be shown, and distances between towns recorded on the map. The heights of mountains were also given. Using these precepts the geographer Chia Tan produced a map in 801 AD which was thirty by thirty-three feet (9 by 10 m) in size. Such an undertaking was far in advance of anything known in the West at that time.

When, at the end of the twelfth century, European scholars began to translate Islamic texts, an interest was reawakened in cartography. Amongst the earliest maps produced in Europe was that of Matthew Paris whose map of England made in about 1250 AD showed little more than the positions of towns, ports and monasteries. A century later the so-called 'Gough' map, made by an unknown cartographer, showed the roads as well, giving fairly accurate distances between locations. Little improvement, however, was made in cartography until the introduction of the clock and magnetic compass, both vital to the accurate fixing of positions. Furthermore, the introduction of copperplate printing in Italy shortly before 1500 AD greatly improved the quality of map production and acted as a spur to the inclusion of greater detail, while the great voyages of discovery created a demand for new and better maps.

R B FLEMING

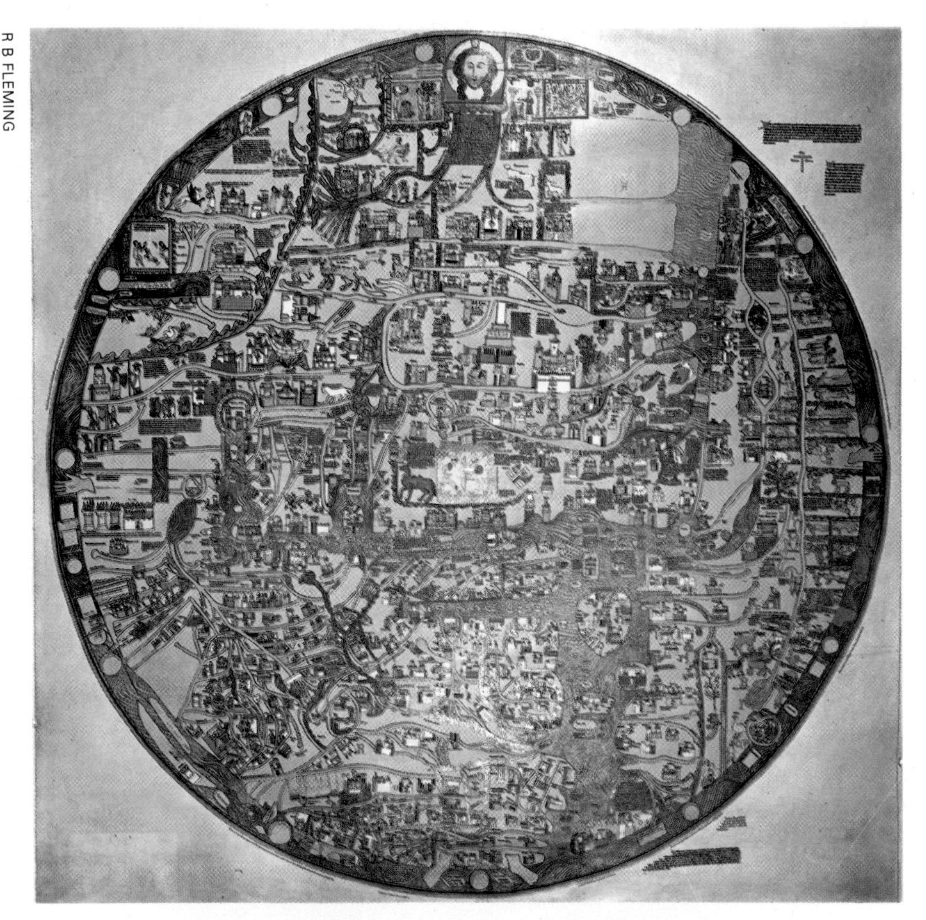

BRITISH MUSEUM

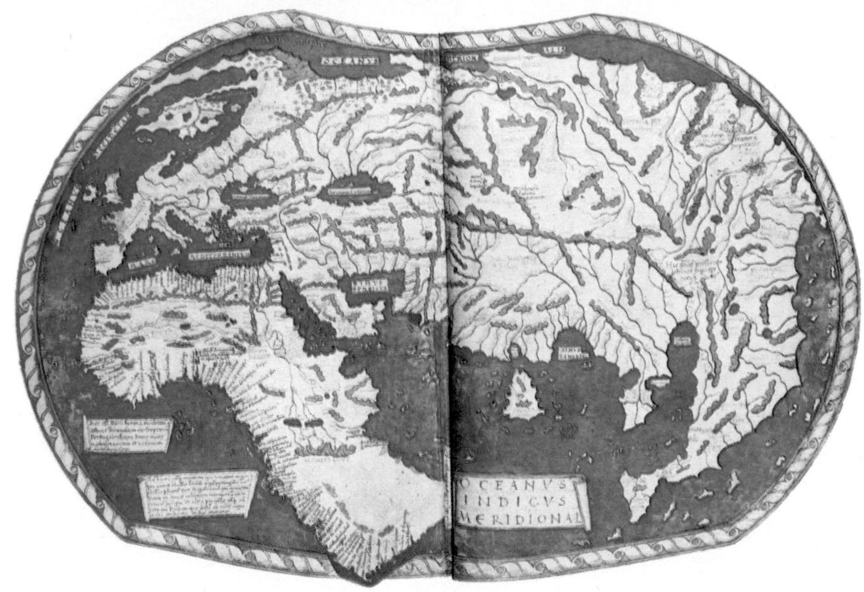

Above left: the Ebstorf world map made in Germany in the 13th century. The map, centred on Jerusalem, shows the Mediterranean sea at the lower right.

Left: the world map of Henricus Martellus Germanus made in 1489 and showing the new discoveries along the coast of Africa.

The water wheel

Two quite distinct types of water-mill appear to have been developed in antiquity, although little is known about their origins apart from the fact that both types were in operation in the eastern Mediterranean and Near East by the first century BC. The Roman architect, Vitruvius, writing in the first century BC gave detailed descriptions of these mills.

The first type of water-mill to be considered is still to be found in operation in remote parts of Scandinavia, the Balkans and countries of the Near East today. The water-wheel was set horizontally below the mill-house, the lower end of its vertical iron axle rotating in a heavy stone bearing. The upper end of the axle passed upwards through a hole in the lower mill-stone and was attached directly to the upper stone. There was, thus, no gearing, and the speed at which the mill-stone turned was the same as that of the water-wheel. The water-wheel was made to turn by water from a fast-flowing stream being directed down a wooden chute against its blades. Commonly the blades of the water-wheel were fixed around its circumference angled to face the oncoming flow of water from the chute, and in this respect resembled some modern turbines. Mills of this kind, however, had serious limitations. They could be used only in mountainous areas where the water in the streams was not only plentiful for a large part of the year but also moved rapidly. Furthermore, because of the lack of gearing the size of the mill, and hence its capacity, was limited.

The main areas of civilization in antiquity, however, were the broad valleys of the large rivers, and here a different, and perhaps more familiar, kind of mill was evolved in which the wheel was set on a horizontal axle to one side of the mill-house. Vitruvius describes two distinct types of such mills. In the one, the so-called undershot mill, the water passed beneath the wheel striking its blades as it did. In the other type, the overshot wheel, water was carried by a chute over the top of the wheel to fill a series of box-like containers set around its circumference. The wheel was thus driven not only by the force of the water but also by the weight of water filling the boxings.

Below: an interesting horizontal mill driven by a water turbine, dating from 1662. The grinding wheels are geared to turn at different speeds.

Bottom: a system of water wheels and pumps designed in 1682 to raise water for the gardens of Versailles, France.

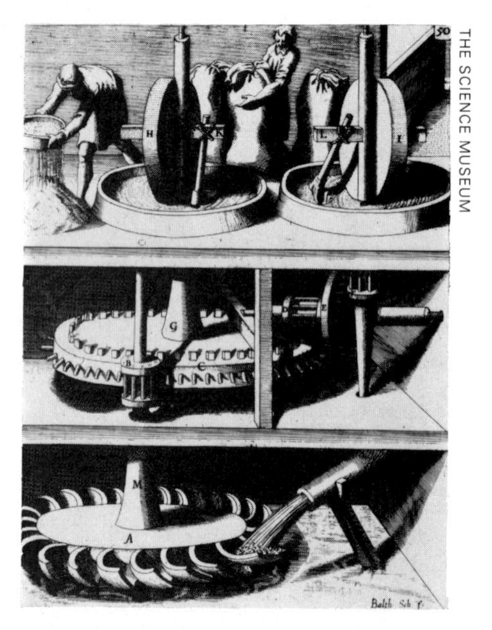

THE SCIENCE MUSEUM

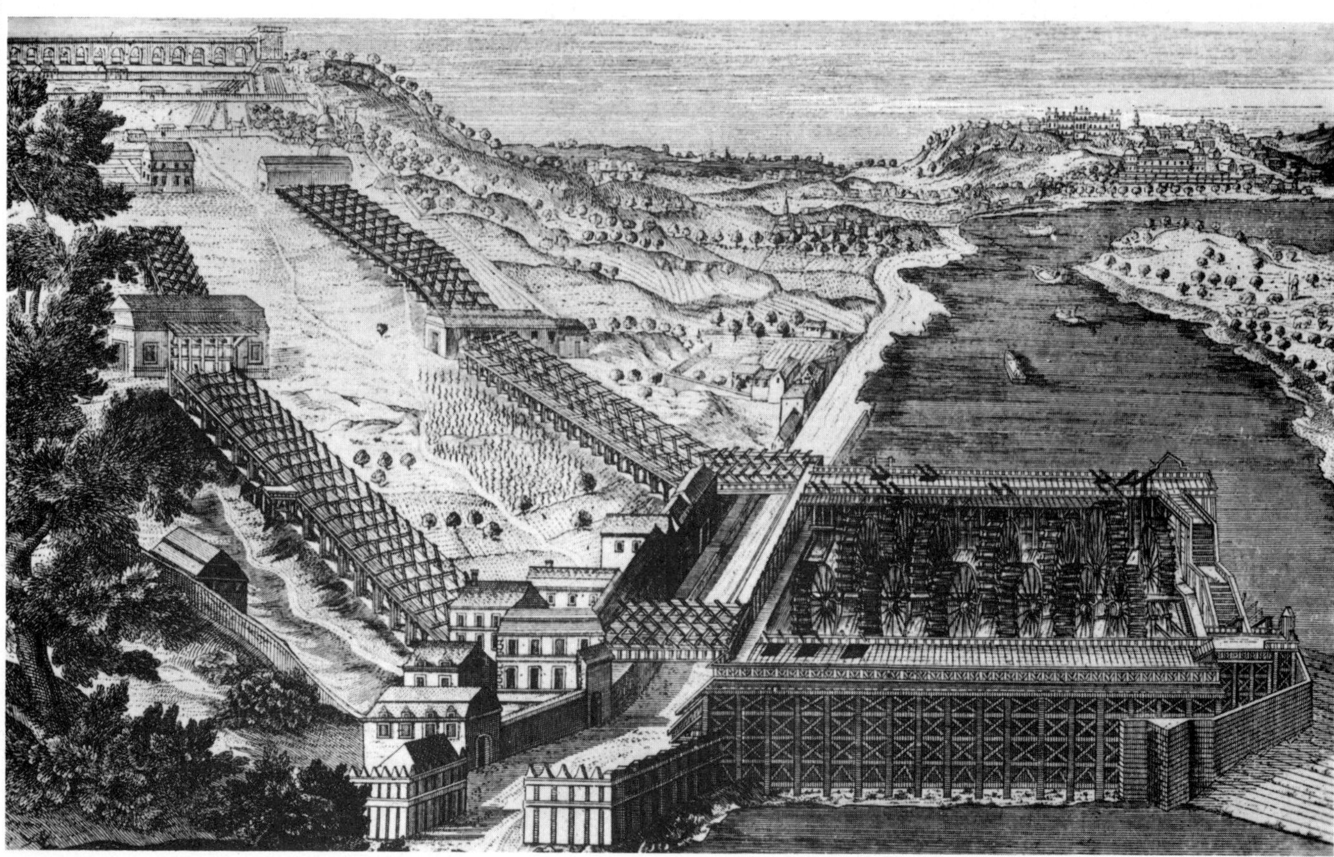

STAATSBIBLIOTHEK BERLIN

The similarity of design between this kind of wheel and the water-lifting wheel used for irrigation is very clear, and there is little doubt that the one influenced the development of the other. Both undershot and overshot wheels had the advantage that a gearing had to be used to turn the mill-stones, and that with correct gearing ratios the mills had a higher output than the simpler turbine type of mill. The overshot mill, however, demanded that water upstream in the river had to be ponded back to provide a sufficient head, and this involved a certain amount of construction work in the form of channels, sluices and dams. The overshot wheel could be used when the river was in spate and in all but the most acute periods of water shortage. The undershot wheel, however, could function only when the river was flowing within fairly restricted limits, for if the water was too high or too low it would not turn at all. One solution to this problem was to construct the mill on a boat's hull or a raft and to moor it in a river so that the flow turned the undershot wheel. Floating mills of this nature are believed to be of late Roman origin, having been designed by the General Belisarius in 537 AD when Rome was under siege by the Goths. Contemporary illustrations of mediaeval towns often show floating mills moored to banks or bridges.

Water-mills were used widely throughout the Roman Empire and in northern China at much the same time. Where the river fell sufficiently rapidly the Romans are known to have built a series of mills, the water discharged from one wheel, being used to turn the next, and so on down the whole incline. The remains of such a range of mills was discovered near Arles, Narbonne, at that time a port supplying both the local military camp and the population of the province. To what extent the Romans used water-mills for other purposes than grinding grain is questionable, but certainly other applications were not common.

During the early part of the Middle Ages the water-mill became a major source of power. The number of mills —as many as 5624—recorded in the Doomsday survey for example is surprisingly high, there being few communities without a mill, and many with two or more. Even in about 800 AD the Emperor Charlemagne found there were sufficient water-mills in France to make their taxation profitable. By 1300 Rouen, then a market town of no great size, could boast of a dozen mills, and this picture could be repeated throughout much of Europe. By the eleventh century, however, it is quite evident that all the mills were not being used to grind corn. Many were adapted to other purposes, and the use of the water-wheel in industry as a source of power opened a new phase in the technological development of mediaeval Europe.

JESPERSEN

Top: the loft of the Slusegaardens Mill in Denmark, built about 1800. The grain was poured in the funnel on to the mill stones inside the casing. These were driven by the system in the lower picture: the largest wheel was turned by the water wheel outside, and the single gear wheel turned the millstones. The side drive operated a circular saw at a nearby farm via a rope on the pulley, driving a smaller pulley on the saw.

The wheelbarrow

The origins and development of the wheelbarrow hold an important place in the history of inventions, not because in itself it represented a major technological innovation, but because it so clearly illustrates a phenomenon known as independent invention. Thus, the wheelbarrow had more than a single source of origin and was developed at different places and at different periods to be used for different purposes.

The earliest form of wheelbarrow was developed in China shortly before AD 200, and traditionally its design is attributed to a general in the Imperial Army, although this is probably mere folklore. The shape of the Chinese wheelbarrow alone is enough to suggest how it originated. The single wheel is large, often three or four feet in diameter, and usually has a dozen or more spokes. It is set between the shafts at a considerable distance from the front end of the barrow, while a boxing or wooden frame is built up from the shafts to encase the upper half of the wheel. Platforms project from the shafts on both sides. The goods to be carried were loaded on to these platforms and on to the upper surface of the boxing. If the burden were correctly placed not only would its weight balance on both sides of the barrow but it would also balance fore and aft of the axle of the wheel. The weight of the load was, hence, almost entirely supported by the wheel, and the man pushing the barrow could use his energy solely for moving it and maintaining its balance.

This kind of barrow appears to have been developed from a two-wheeled handcart in which, with modifications, the pair of wheels was replaced by the single central wheel. In many parts of China, especially in the wet, rice growing areas, field boundaries are often no more than narrow embankments, the tops of which must also serve as paths. This wheelbarrow, therefore, seems to be an adaptation of the handcart designed to allow it to move along these narrow paths. Indeed early illustrations of Chinese wheelbarrows invariably show them being used to carry vegetables and other agricultural products.

In Europe the wheelbarrow was clearly developed from a type of hod carried by two men. In this device a wooden box to hold the load was set between a pair of long handles projecting in front and behind, one man supporting the front end, the other supporting the rear. Four stumpy legs were often provided so that when the men came to lift the hod, the handles were clear of the ground. By replacing the leading carrier by a small wheel, often with no more than four spokes, a form of wheelbarrow was created. In some illuminated manuscripts both the two-man hod and the early wheelbarrow are shown together, making the derivation of the one from the other perfectly clear.

The European wheelbarrow seems to have been designed in the twelfth century, and all the early illustrations show it in use on building sites, where it is being used to move stones and mor-

Below: in his great textbook on mining, 'De Re Metallica' (1556), Agricola gave much information on various tools including this early wheelbarrow without legs.

Right: wheelbarrows were also used widely in building, as seen in this 15th century Flemish manuscript.

JOHN FREEMAN

RADIO TIMES HULTON PICTURE LIBRARY / SALLY SLIGHT

tar. In this respect its function was quite different from that of the Chinese wheelbarrow, and indeed the forward placing of the wheel meant that the men using the European wheelbarrow had to lift a large part of the burden, quite apart from pushing and balancing. It was thus totally unsuitable for shifting goods over long distances.

In recent centuries Chinese barrowmen, who often travel long distances, have often been seen to use sails to help them along the road, although this can only be done when the wind blows from behind them. Usually a pair of poles are lashed vertically to the barrow to serve as masts from which a square sail of cloth is set, and it is believed that this practice is almost as old as the wheelbarrow itself.

It is a curious fact that the Chinese never developed a wheelbarrow comparable to that used in Europe which, despite its mechanical inefficiency was often more convenient to use. The Chinese would often lash a load between a pair of bamboo shafts to be carried by two bearers. Furthermore, by the eighth century they had developed a type of sedan chair borne in the same manner, yet at no stage did it occur to them to replace the leading carrier by a wheel. Equally, handcarts were known in western Asia and Europe certainly by 1000 BC, and there are pictures of them in use, for example, by the Assyrian army in about 600 BC. It would seem, however, that the need never arose, as it did in China, to restrict the track so severely that a form of wheelbarrow was evolved. Thus, until the seventeenth century, when direct contacts were made between Europe and China as a result of trade, each area had its own distinct form of wheelbarrow, although by this time both were frequently modified to serve other needs than those for which they were originally designed.

Metal ore being transported by wheelbarrow to a furnace where it is transferred to baskets and dumped in the top, the molten metal being tapped from below.

The windmill

The first windmills to be recorded come from Sistan in eastern Persia where, according to tradition, they were invented by a slave, Abu Lulua, in about 650 AD, as the outcome of a wager. Lulua boasted that he could harness the wind and the caliph held him to his word. The windmill that he is said to have invented is of a type that is still to be seen operating in Sistan today.

The building is a tall tower of sun-dried brick. Inside it the vanes that catch the wind, and which are usually made of reed matting, are attached to a vertical windshaft that runs up the centre of the building. This shaft is connected directly to the upper millstone, which it turns. Two apertures are cut into the tower, one to allow the prevailing wind to impinge upon the vanes on one side of the shaft, and the other to allow it to escape; the vanes on the opposite side of the building are protected from the wind. This type of mill can only function in a region where there is a steady prevailing wind always from the same quarter, since its working depends upon the apertures being so orientated that the wind will pass in through one and out through the other. Thus it has never found much favour outside Persia, where a steady wind blows for three months of the year, the so-called 'wind of one hundred days'.

Even if Lulua did not invent this type of windmill, by 950 AD it must have been quite common in Persia, for it is mentioned at this time by two Islamic geographers. The first known windmills in western Europe appeared two centuries after this date and were so different in design from the Persian ones that they must have been invented independently.

The earliest European windmills took the form of a small wooden millhouse with a gabled roof, which was supported on a single, stout wooden post about which it could rotate. For this reason it is usually referred to as a post-mill. The post was held rigid in the ground by cross-beams and ties. On its upper end was an iron bearing that was engaged in an equally sturdy cross-beam inside the millhouse. The rotor was made of four lattice sweeps that were covered with cloth, the so-called sails, and these were mortised into the windshaft. The windshaft, canted at an angle of about ten degrees from the horizontal, rotated through bearings in the front and rear timbers of the millhouse. The teeth of a wooden crown-wheel on the windshaft engaged between the spokes of a cage-wheel, the axle of which drove the upper millstone. There was, thus, a gearing between windshaft and millstone, and this was usually of a ratio that allowed very rapid movement of the stone. The millhouse could be turned into the wind, no matter from which quarter it came. To do this the early mills were equipped with a long pole, the tail post, which stretched from the floor of the house to the ground where it could be lashed to a post to hold the mill in position.

From 1180 onwards post-mills were to become a very common sight throughout western Europe, being built not only on hilltops in open countryside, but also on the walls of cities and castles. The Crusaders, for example, are known to have built windmills on the walls of their fortifications. The windmill clearly had many advantages over the older watermill. A windmill could be

Below: the post-mill was supported on a stout post around which it was turned to face the wind. The tail post was secured to hold the millhouse in position.

BODLEIAN LIBRARY

Below: the fantail, a mid 18th century development, automatically turned the sails into the wind. To reach the sails, a stage was built around the tower.

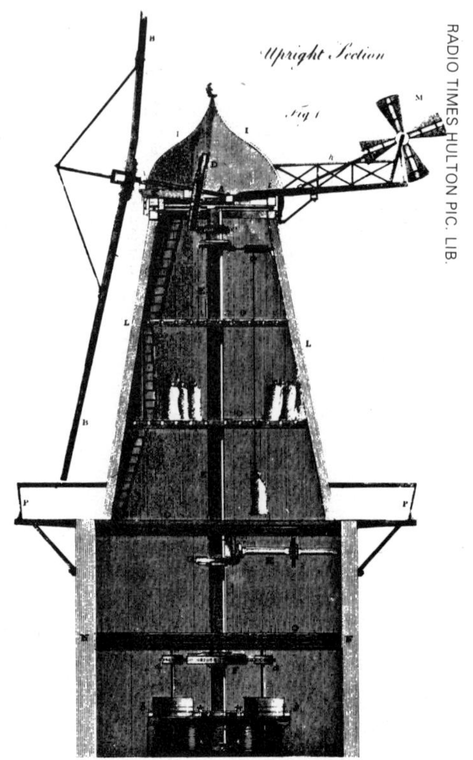

RADIO TIMES HULTON PIC. LIB.

ROBERT HARDING ASSOCIATES

constructed away from any river and, in the depth of winter, it was not dependent upon rivers remaining unfrozen.

Curiously, we know almost nothing about the precise origins of the post-mill. Since, however, the post-mill was a wooden structure and all early references to it come from northern France and England where, in the twelfth century, most domestic buildings were made of wood, it is likely that it originated in these countries. Also at this time the monastic orders, especially the Cistercians, were showing increasing interest in many forms of engineering, so that the design of the post-mill may well have been the work of a group of monks.

One other type of windmill which appeared in the Middle Ages is the tower-mill, so called because it is usually a stumpy, stone tower. It has a low, conical roof into which is set the wind-shaft carrying the sweeps, and it is this roof alone that can be turned into wind. This type of mill can still be seen in more or less its original form throughout the northern Mediterranean area, from Spain to Cyprus. It was probably an adaptation of the post-mill to suit these countries, which lack the heavy timbers required to build the original form. Although today most of these mills carry small, triangular cloth sails, in the late Middle Ages they were fitted with lattice sweeps typical of the European post-mill, such as those still used on tower-mills in Malta and Gozo.

Above: the sails inside these Afghan windmills are attached to a vertical wind shaft and made of reed matting.

Below: the stumpy, stone tower-mill with small triangular cloth sails is typical of the northern Mediterranean region.

ROBERT ESTALL

Nets and rope

It is very difficult to say even approximately when early man began to use either ropes or nets, partly because the materials of which they were made are apt to decay rapidly unless buried under ideal conditions, and partly because the tools used in their manufacture were of such a simple kind that they might have been used for many other purposes. Today when we speak of ropes and nets we normally think of them as being made of plant fibres such as jute or hemp, if not nylon, but in the remote past many other materials may have been used, such as sinews, thongs of hide, or even hairs.

The first hint that mankind was using nets comes from the Old Stone Age, the prehistoric period that ended about 10,000 BC, when man lived mainly by hunting large animals and by fishing. Painted on the walls of the caves in which he dwelt in the south of France are a number of criss-cross patterns which may well be intended to be nets, while among the rubbish unearthed from the cave floors have come a number of bone needles that could have been used in the manufacture of netting. Thus, although the evidence for the use of nets at this stage is far from conclusive, the Old Stone Age man was a very competent toolmaker, and with the pattern of the spider's web to guide him the probability remains that he commonly used nets to catch his game and fish.

In the following period, the Middle Stone Age in Europe, which lasted until about 5000 BC, man certainly used fishing nets, for small fragments have been recovered from peat deposits in which they have been quite well preserved. Furthermore, a number of small stones with holes bored through their centres have come to light, and these were probably used as sinkers for fishing nets. The nets would have been either the circular type, thrown over a school of fish, or the rectangular sort that were suspended across a river.

The most complete picture for the use of netting in the prehistoric world comes from Egypt from about 5000 BC onwards. Materials of this kind were extremely well preserved in the dry desert soils, and the Ancient Egyptians also left behind them in their painted tombs a precise record of their daily life. Nets of many different varieties were used not only to catch fish but also to snare water fowl, and even larger game such as lions. Egypt was, however, not

BRITISH MUSEUM/PHOTO: MICHAEL HOLFORD

BRITISH MUSEUM/PHOTO: MICHAEL HOLFORD

Two uses of ropes before 1000 BC.
Above: an Assyrian basrelief showing slaves pulling heavy ropes, driven on by an overseer with a whip.

Left: an Egyptian tomb wall painting dating from the reign of the 18th Dynasty king Tutmosis IV. It shows a boat rowed by slaves being used to tow another boat with court ladies in it.

Both these uses demanded large, strong ropes, showing that even at this early date Egypt and Assyria had quite a developed ropemaking industry, making the best use of available materials.

unique in this use of nets, for the splendid carvings in low relief from Mesopotamia (present day Iraq), although much later, tell the same story of nets being used not only to catch game but also to capture animals to stock the royal menageries.

The use of ropes in the Old and Middle Stone Ages is far less likely than that of nets. While it is true that these early hunters used harpoons in which the barbed head, made of bone, was attached to the shaft by a length of cord, it is likely that the cord was simply a thong cut from hide. Similarly, lassoos and noose-traps were probably not made of rope. Indeed ropes appear to have been developed in the first instance as an essential part of the equipment of boats in the early part of the New Stone Age. Perhaps it was the need to find some material that did not become hard and inflexible after it became wet that led to this development.

In Ancient Egypt ropes were employed as hawsers (anchor ropes) and as rigging for masts and sails; since the normal Nile boat was made by lashing together bundles of papyrus reed, ropes were also used in their construction.

In both Egypt and Mesopotamia the papyrus reed—the bulrush of the Bible—provided the raw material from which ropes were made. The leaves were first cut into narrow strips which were then spun, as one would spin wool or flax, to give threads. A number of threads were then laid together and twisted to form strands; finally the strands were laid together and twisted to form the rope. An Egyptian tomb-painting of about 2000 BC shows an early rope-walk in which the rope-maker is twisting the strands together by means of a wooden rod passed between them.

Although ropes were first used for boats and shipping in both Egypt and Mesopotamia they became essential in the craft of building. The huge blocks of stone used to construct the Pyramids and the vast statues that adorned the temples could only be moved by men hauling on ropes. A fragment of such a rope from the ancient Egyptian city of Saqqara was found to have a diameter of six inches (15 cm) and could, when new, have withstood a pull of ten tons.

One might imagine that ropes were not used in prehistoric Europe lacking as it did any large boats or stone buildings. Evidence suggests otherwise. Near the bottom of a deep well of Bronze Age date (about 1500 BC) in Wiltshire was found a section of rope made from the fibres that underlie the bark of the lime tree. It was presumably a part of the hoisting equipment for the well. The birch, lime and willow all yield suitable fibres from which to make ropes, and until imported fibres such as jute and hemp became common in the eighteenth century the majority of rope in Europe was made of these materials.

SCALA INTERNATIONAL

ZEFA

Above: this Roman mosaic pavement, though not itself particularly ancient, shows one of the earliest uses of nets: for hunting animals. The only later feature is the horses ridden by the hunters. The net is weighted at the bottom to prevent animals from running under it.and slung between two tree stumps. The animals are then driven into it. They are caught not only because the net is a barrier but because they catch their feet—and in this case antlers—in the meshes.

Left: an Inca suspension bridge over the Rio Pampas in Ayacucho, Peru. These shaky looking structures are in fact stronger than they appear. The crudely twisted cords, made of plant fibres, are not particularly strong in themselves, but there are so many of them that the total strength of the bridge is fairly high.

Boats

Unlike many of man's early inventions there was no single prototype for the boat. Instead boats developed from a number of different primitive water craft in different places and at different times. Basically all boats were developed from either floats, rafts or watertight baskets, many of which are still used by primitive peoples today.

A log may be used, for example, as a float to help one cross a river. This can be done simply by sitting astride the log and paddling with the hands. If the log is hollowed out by means of an axe and fire it is possible to not only keep dry but also carry a load. The dugout canoe in one form or another is known in every continent of the world, and it seems likely, therefore, that this invention alone has been made several times. In Europe and Asia dugout canoes are known from the time that stone axes became widespread in the New Stone Age from about 6000 BC onwards. The simple dugout canoe, however, tends to be unstable and will easily capsize.

To prevent this happening an outrigger float was added, or two canoes might be lashed side by side, and such craft were common until quite recently in the Indian Ocean, South East Asia and the Pacific. An alternative solution was either to shape the hull, so giving it a certain amount of 'keel', as is still done in tropical Africa and South America; or to place weights in the bottom of the craft, which seems to have been the common practice in prehistoric Europe.

The dugout canoe has very little freeboard—height above the water—and to overcome this fault an additional plank, a *washboard*, may be added to each side of the vessel, the joint being made watertight with resin or BITUMEN. Indeed, a number of such planks might be added in this way one above another with a row of vertical supports, or ribs, placed on the inside for increased strength. Early pictures of Cretan boats, from about 2500 BC onwards, look as though they were constructed in this way.

In Egypt, where large timbers were not available, the earliest boats were made by lashing together a number of cigar-shaped bundles of reeds, and some streamlining was given to the vessel by pulling the two ends up from the water by means of ropes. Paddles were used to move them through the water and steering oars kept them on course. When the Egyptians wanted to build a sturdier vessel for the transportation of heavy stone, although they kept the shape of the reed raft, they used the plank construction already described. In Africa today reed rafts like those of ancient Egypt are still in use in Abyssinia and on Lake Chad, while very similar boats are also used in Peru.

In early China flat bamboo rafts were in use probably from the beginning of the New Stone Age, about 4000 BC, and as a cheap means of river transport are still in use today. At a later date, perhaps about 1000 BC, the Chinese began to adapt the raft by laying the bamboos along the curved sides of a series of semi-circular wooden planks, thus creating a vessel that had a number of solid bulkheads down its length. Finally, by replacing the bamboos with planks the Chinese created their typical river and harbour craft, the wupan (five planks) and sampan (three planks). A clay model of such a boat was found in a Han tomb of about 100 AD. The Chinese junk was developed from this type of vessel.

The third type of boat, the wooden frame covered with skins or bark, and made watertight with bitumen or resin survives today in the form of the Irish curragh, the Eskimo kayak and the American Indian canoe. The people of the Mesopotamian delta were making similar craft by 3000 BC, as we know from both pictures and models, while in Scandinavia prehistoric engravings of about 1000 BC show that there, too, frame boats were normal. Indeed the Viking long boat did not begin to be developed until the fourth century AD and was no more than a clever adap-

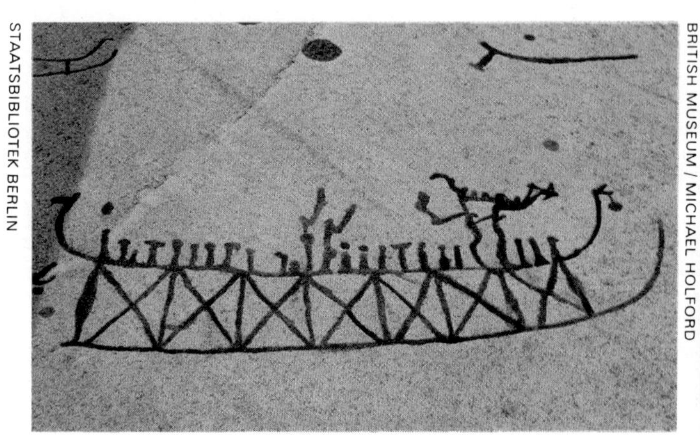

STAATSBIBLIOTEK BERLIN

Above: Bronze Age cave drawing of about 1000 BC *at Tanum, Sweden, showing a type of wooden framed, skin covered boat.*

BRITISH MUSEUM / MICHAEL HOLFORD

Right: detail from a relief found at Nineveh, Mesopotamia, describing the expedition of King Sennacherib to Phoenicia. It depicts a Phoenician war galley of about 700 BC, *powered by rows of oarsmen. These ships had a long pointed prow, set low down, which was used to ram and sink boats.*

tation of the rib-and-plank construction pioneered in the Mediterranean.

Primitive people either tow, pole or paddle their boats, and so it was with the very early craft. The oar, which allows better use of manpower, seems to have been developed in the Mediterranean somewhere about 3000 BC, and was soon adopted by the Egyptians. Its use did not spread to some areas, such as the Pacific, until the eighteenth century.

Sails are shown on Egyptian river craft as early as 4000 BC. These were square sails set upon a horizontal yard arm. The square sail was to become the common form not only in the Mediterranean and the west, but also in China. Shortly after 200 AD the Chinese discovered that by sewing a number of horizontal battens into the sail it could be made to lie in line with the ship and thus allowed SAILING very close to the wind. The triangular fore-and-aft sail favoured by most yachtsmen today was, however, an Arab invention and did not begin to come into use in Western Europe until after the Crusades.

Originally steering was achieved by suspending an oar over one or both sides of the craft at the stern. Shortly before 200 AD the Chinese realised that a better solution lay in setting the oar vertically through a shaft in the overhanging deck of their boats, and the RUDDER continued to be used in this form on all Chinese junks. It was probably as a result of contact with the Chinese that Arab seamen began to hang the rudders like a hinged door on to the stern-post of their ships some time before 1000 AD. A century later, as a result of contact with the Arabs during the Crusades, the sailors of Western Europe began to use the rudder.

ZEFA

ZEFA

KEYSTONE

ASPECT

Above left: fishermen on the coast of Orissa, India, use boats which are made by lashing together several roughly shaped planks to form a sort of canoe.

Above right: the simple dugout canoe, known to date back about 8000 years, is still in use in many parts of the world today. These are canoes of the lagoon fishermen of Dahomey, in west Africa.

Far left: a 'balsa', a boat made from bundles of reeds bound together, on Lake Titicaca in Bolivia. The design of these boats is very similar to that of the reed boats of ancient Egypt.

Left: the coracle, made from skins or tarred canvas stretched over a wicker frame, was first used by the early Britons and can still be found in some areas. These are on the rivers Teifi in Wales.

Sailing

BRITISH MUSEUM/PHOTO: MICHAEL HOLFORD

AMPLIACIONES Y REPRODUCCIONES MAS

The first boats known to have been fitted with sails were those of Ancient Egypt of about 3000 BC, seen in crude paintings on pottery of the period. Possibly sails were in use even earlier, because as shown by Thor Heyerdahl in his voyage on the *Kon-Tiki,* simple rafts are quite easily sailed. Like the *Kon-Tiki,* early Egyptian sails were set on a mast made of two poles lashed together at the top and spread at the base, a system that was essential on a boat built of reeds. The sail itself was a square sheet of papyrus matting or of linen attached by its upper edge to a single spar *yard-arm. Halyards,* ropes running to the top of the mast and down again, allowed the yard-arm to be hoisted when the sail was needed.

By 2000 BC square sails were in use on boats throughout the whole of the eastern Mediterranean. The vessels were made of wood which allowed a single pole mast to be used, unlike the two pole system of reed boats. By 1500 BC sailors in both Egypt and the Mediterranean had begun to attach the foot of the sail to a horizontal boom, and by *reefing* (rolling up or folding and tying) the sail to the boom as much or as little of the sail could be used as the wind demanded. This system was not, however, to last for long, because by 1200 BC sails were generally fitted with *sheet-lines,* cords running vertically through the fabric of the sail. By hauling on the sheet-lines the sail could be furled to the yard-arm, while by paying them out partly or wholly the sail could be adjusted to the weather. This square sail with a free foot and sheet-lines was to become the universal sail of the Greek and Roman world. The sail was generally made of linen, although Caesar found the Gaulish sailors on the Atlantic coast using leather sails. Normally ships carried only a single square sail, but after 200 AD Roman ships are often shown with an additional sail, the *artemon,* carried at the bow.

Top left: this Egyptian buff decorated pot, characteristic of the Naqara II or Gerzean pre-dynastic culture (3300 BC), depicts a boat with sail and oars.

Left: a Roman mosaic floor showing Roman trading ships with square sails supported on a single yard arm with no lower boom.

The square sail had one great disadvantage: it could be used only in a following wind. If the wind blew from ahead, or even from the beam, it was impossible to sail, and for this reason ships fitted with both sails and oars remained common in the Mediterranean. By about 100 AD, however, Arab sailors had invented a new type of sail that overcame this difficulty.

In the few centuries before 100 AD the Arabs had been sailing from Aden, Oman and the Persian Gulf to East Africa, India and Ceylon, trading in spices, ivory and slaves. To travel quickly from Arabia to Ceylon in the monsoons they needed a sail that would carry them with the wind on the beam. They found the answer in a triangular sail attached along its upper edge to a long sloping yard-arm. Such a sail could be trimmed to lie fore-and-aft, down the length of the vessel, so catching the wind from the beam. By 500 AD the sailors of the Mediterranean had adopted this Arab sail and it became known in the west as the *lateen* (Latin) sail.

Seamen of the Atlantic coast were slow to adopt the lateen sail, and for most of the Middle Ages their ships carried square sails. By 1400 AD, however, many sailors were using lateen sails, as in the Portuguese *caravel* (a small fast sailing ship with three masts), or a combination of both square and lateen sails. The square sail was usually on the main mast and the lateen on a mizzen (rear) mast, an arrangement that was continued in the great ships of the 16th century, although the number of sails was increased. This type of rig was used by Vasco da Gama when he rounded the Cape of Good Hope and by Columbus when he discovered the American continent.

The lateen sail, however, was not the only answer to sailing with the wind ahead or on the beam. By 200 BC the Chinese were using a square sail similar in most respects to those in the West, but they were also using a matting sail which was stiffer than a cloth sail and more capable of being trimmed fore-and-aft. The logical development from this, first seen in about 100 AD, was a cloth or matting sail fitted with a number of horizontal battens and a boom at its foot. This was the forerunner of the lug-sail of south east Asia, and when Europeans first came across shipping in this area, from about 1600 AD onwards, most vessels were propelled by such lug-sails, either of plain cloth or with battens. Battened lug-sails, often set on three or four masts, were to become the standard rig for Chinese sea-going ships, the junks, which until 1433 were sailing as far afield as the east coast of Africa.

Below left: a fishing boat from Bali, Indonesia. The sail is mounted on fork shaped masts and the boat is stabilized by wooden outriggers, which act like keels.

Bottom left: Nile boats with lateen sails—triangular in shape, attached to a yard arm.

Below: a Chinese junk in Hong Kong harbour. The sails are stiffened with battens.

ASPECT PICTURE LIBRARY

ZEFA

ASPECT PICTURE LIBRARY

The rudder

All early ships were steered by oars suspended over the sides of the vessels at their sterns, although the number of oars used, and the way in which they were fixed, varied considerably. Thus the Ancient Egyptians commonly employed one, two, or even three oars on either side of the vessel, while the Vikings used a single oar attached to the starboard (steer-board) side of their longships. These steering oars had many disadvantages. Projecting over the side of the ship, they had to be raised when docking; to be efficient they often had to be of a deeper draught than the keel; they caused considerable drag; and they were not very effective unless, as in Ancient Egypt, used in large numbers.

The first shipwrights to introduce a rudder, suspended at the stern of the vessel in line with its axis, were the Chinese. By 200 AD the design of Chinese shipping had evolved to a point where the upper deck overhung the stern of the vessel to provide a *transom*, and it was a relatively simple matter to make a vertical perforation through this deck through which the steering oar could be suspended. At the same time the blade of the oar could be enlarged to become more effective without creating appreciably greater drag. This type of rudder was suspended in its socket by ropes since the design of the hull was such that there was no *stern post* to which it could be hinged, and until recently the traditional junk has preserved this style of rudder.

By 850 AD merchants from southern Arabia and the Persian Gulf were regularly making the voyage to Canton and back, and during their stay in China they must have encountered vessels fitted with a rudder. Presumably they realized the advantages that it offered, but the problem that they faced was that it had to be adapted to suit their own ships. Their vessels had no transom, the hull being formed on to a steeply rising post projecting high above the *gunwales* (upper edges of the sides). The ultimate solution lay in hinging the rudder to this post. The tiller was in the form of a cross-bar attached to the rudder, and from its ends ran ropes connected to pulleys on either side of the ship. The rudder was thus controlled by the helmsman heaving on one or other of these ropes, an arrangement

SCALA

Above: a Byzantine mosaic in Ravenna, Italy, dating from the sixth century, and showing ships with steering oars.

Below: this carving in Winchester Cathedral, Hampshire, England, shows an early European stern-post rudder.

PETER SCHMITT

MICHAEL HOLFORD LIBRARY

Above: the Incas of Peru steered their rafts by using a board, such as this, at each corner to create variable drag.

Below: this engraving dating from about 1800 shows the efficient use of a stern-post rudder in China.

still to be seen today in many of the traditional forms of Arab craft.

Unfortunately Islamic writers and artists have left us few accounts or pictures of shipping. There is, however, a picture of a vessel on a Persian plate of about 1000 AD which proves that by that time the rudder was in use by at least a few Islamic seamen. By the thirteenth century there is a sufficiently large number of pictures of sailing craft from Islamic countries for us to be certain of the details. Even so, at this later period there were still many ships controlled by steering oars.

By 1200 AD some western European sailing craft were beginning to be fitted with stern-post rudders, there being an illustration of such a ship carved on the font of Winchester Cathedral, which is of about this date. Unlike the rudders in Islamic countries, however, those in Europe were usually fitted with bar tillers attached by a short cross-piece or directly to the rudder's stock. It is possible that these European rudders were copied from Venetian and Byzantine merchants in the eastern Mediterranean who in turn had adopted the rudder from the Arabs. In this case the rudder would have been one of the ideas brought back to western Europe by the crusaders. On the other hand, it has to be remembered that Scandinavian boatmen frequently made the passage from the Baltic to the Black Sea by way of the Dvina and Dniepr Rivers, and in the Black Sea they certainly came across Islamic craft. In fact some of the earliest pictures of ships fitted with stern post rudders in Europe appear on the seals of Baltic ports such as Lübeck and Elbing in the thirteenth century, so that it would seem just as probable that other Europeans learnt of the rudder through the Scandinavians by way of the Crusades.

The introduction of the stern post rudder had a considerable effect on the design of European shipping. The high stern-post, a feature of all European ships since the Viking period, had to be cut down to make room for the tiller. This in turn allowed the reshaping of the poop and aftercastle, often to give a transom stern curiously like the Chinese vessels in which the rudder originated. Such ships, seaworthy and manœuvrable, made the voyages of discovery at the end of the fifteenth century a practical matter hardly possible with earlier forms of sailing craft.

MARY EVANS PICTURE LIBRARY

ECONOMY OF TIME AND LABOUR EXEMPLIFIED IN A CHINESE WATERMAN.

Pulleys and gears

The pulley-wheel, grooved to take a rope or cord, seems such a simple device that one might expect it to be almost as old as the wheel itself. In fact the pulley appeared quite late in man's history, and there is considerable doubt as to when and where it was invented. What may be taken for a pulley appears on a low relief of the Assyrian period (about 800 BC) in Mesopotamia, while an actual pulley-wheel of a slightly later date made of mulberry wood was found in the ancient Mesopotamian city of Nimrud. These pulleys were probably used for raising water from wells. On the other hand, pictures of Greek ships of about 600 BC show that the hoists for the yardarms were carried through rings attached to the tops of the masts, and not as one might expect through pulley blocks. Evidence exists, however, that pulleys were used in shipbuilding somewhat earlier.

In ancient Greek drama during the fourth century BC, pulleys and ropes were used to lower a god character to the stage from above, as if from heaven, in order to solve the problems of the mortal characters below. (Hence the phrase *deus ex machina,* or the god from a machine, used in drama today to mean a plot-resolving device.)

Simple pulleys were, however, known by the time of Aristotle (about 325 BC) since they are described in a book, *Mechanica,* written by one of his contemporaries. Compound pulleys, in which a rope or cord passes over two or more pulley-wheels, are said to have been invented by Archimedes of Syracuse who died in 212 BC. It is doubtful if Archimedes really invented compound pulleys, but he certainly made dramatic use of them, on one occasion pulling a galley up on to the beach single-handed with a combination of ropes and compound pulleys.

The Roman writer, Vitruvius, who lived in the first century BC, and Hero of Alexandria, who worked in the first century AD, have left in their books clear accounts of compound pulleys, and it is evident that they fully understood the mechanical advantage to be gained by using combinations of large and small pulley-wheels.

Illustrations of ships, cranes, oil-presses and other machines from the same period as that in which Vitruvius and Hero were working show that pulley-blocks had become a common feature throughout the Roman Empire. They are often depicted being used in combination with a capstan, a large wheel mounted on a vertical axle, or a windlass, a similar wheel mounted on a horizontal axle. Usually they were worked manually but by attaching pedals to the windlass it became a tread-wheel, thus using the whole of the operator's weight to turn it. By

Below: an early fourteenth century picture of a pulley used with a windlass for hoisting. The windlass lets the operator lift weights heavier than himself.

Below right: simple wooden pulley blocks in use at an Indian well. Large buckets of water are hauled up in this way by oxen (out of the picture to the left).

MARY EVANS PICTURE LIBRARY

MARGARET MURRAY

attaching a horizontal shaft to the capstan and yoking an ox or an ass to it, an animal could work the capstan by walking a circular path around it. The windlass and capstan were described by both Vitruvius and Hero.

Curiously in antiquity these devices were never used with a cord or belt drive. Instead, where gearing was necessary, wooden toothed wheels were employed. Again, Archimedes is credited with having invented the toothed wheel, but it seems more likely that he merely popularized the device.

The importance of the toothed wheel lay in the fact that by using combinations of wheels of small and large diameters one could arrive at a suitable speed or work ratio for any machine. The principles of this kind of gearing were clearly fully understood by Vitruvius and Hero. Vitruvius for example described how a capstan, made to turn by an ox, could be used to operate a water-raising wheel. Teeth set around the circumference of the capstan engaged with those of a smaller wheel fixed to a horizontal axle that drove the water-raising wheel. Thus the slowly moving capstan worked by the ox could be made to operate the water-raising wheel at a relatively higher speed.

Hero was to carry the development of toothed-wheel gearing even further in his *hodometer*. This was a machine in which distances could be measured directly according to the number of revolutions made by the wheel of a cart.

To achieve this Hero used a complex arrangement of gears so that one could read off the exact distance covered at any time.

Machines using pulleys, capstans and toothed-wheel gearing remained in use after the decline of Rome, and these elements are to be found, for example, in mediaeval cranes, water mills and windmills, while translated into metal the toothed wheel was to become an essential part of the clock.

Early gear wheels used wooden pegs as teeth. They were often extremely large. This set is at a salt mine at Wieliczka, Poland (now a museum) and probably dates from the 18th century.

ERICH LESSING MAGNUM PHOTOS

Glass

In Mesopotamia shortly before 2000 BC it was discovered that a mixture of silica sand and an alkali such as soda or potash, when heated strongly, would fuse and liquefy to form a glass. Long before this, however, craftsmen throughout the Near East had been manufacturing another material made of the same ingredients, generally known as Egyptian faience. It was used to form amulets and small vessels, the sand and alkali mixture being heated only sufficiently to allow the sand grains to fuse and a glassy film to form on the surface. There seems little doubt, therefore, that glass was first discovered as the result of over-heating Egyptian faience.

Glass objects were rare and valuable until 1500 BC. Formulas for glass making are found on Mesopotamian cuneiform tablets, but curiously, although glass making was an important industry in ancient Egypt, little written reference to it has been found there, and it is not known whether glass vessels originated there or elsewhere.

Initially the Mesopotamian craftsmen, having formed the glass, allowed it to cool and then shaped it with abrasion tools as though it were a piece of stone, a technique long established in this area for the production of amulets and seals. It was soon discovered, however, that glass was more easily shaped while hot and plastic. Vessels were made by one of two methods. Either a core of sandy clay was built up on the end of a rod and this was dipped into the molten glass and withdrawn, the sandy core being scraped away after cooling; or lengths of glass rod were first produced. These were then reheated and coiled around a core, and finally heated again and rolled to and fro on a stone slab to smooth the surface of the vessel.

At first little attempt was made to produce a clear colourless glass. The raw materials normally contained many impurities. The presence of iron compounds rendered the glass green or amber, while other impurities made it opaque. Compounds containing copper were frequently added to the molten glass to make it blue, and often threads of white or amber glass were arranged on the surface of vessels and then rolled into it after reheating.

These cumbersome methods of shaping glass remained in use until the second century BC, when glass blowing first began to be practised in the countries of the Levant. Indeed, it was not until blacksmiths had found a means of shaping a long iron tube that glass could be gathered on the end of an iron tube, blown into a thin-walled balloon and, with regular reheatings, further worked into the required shape. Glass working, until then the monopoly of the Middle East, rapidly spread throughout the Roman Empire.

As an alternative to the free blowing of glass, some workers preferred to use moulds, the gathering of glass being blown into a piece-mould which carried an imprint of the design of the glass vessel being made. Furthermore, with the introduction of glass blowing techniques translucency became a possible and much desired feature of glass vessels. Great care was taken in the choice and preparation of the ingredients of glass to eliminate impurities, especially iron compounds, with the aim of making a clear, water-white material. Equally, cobalt and manga-

BRITISH MUSEUM/MICHAEL HOLFORD LIBRARY

Left: a decoration of Egyptian faience, which was widely used for making small amulets, beads and flasks at an early date.

Below left: a vessel of sand core glass, made by coiling molten glass rod around the core. The pattern was produced by combing a point down the coils. This piece dates from the 6th to 4th centuries BC.

Below: a Syrian flask in the shape of a head, made from blown glass, with colouring material added in a random way for effect. It dates from 200 to 300 AD.

BRITISH MUSEUM/MICHAEL HOLFORD LIBRARY

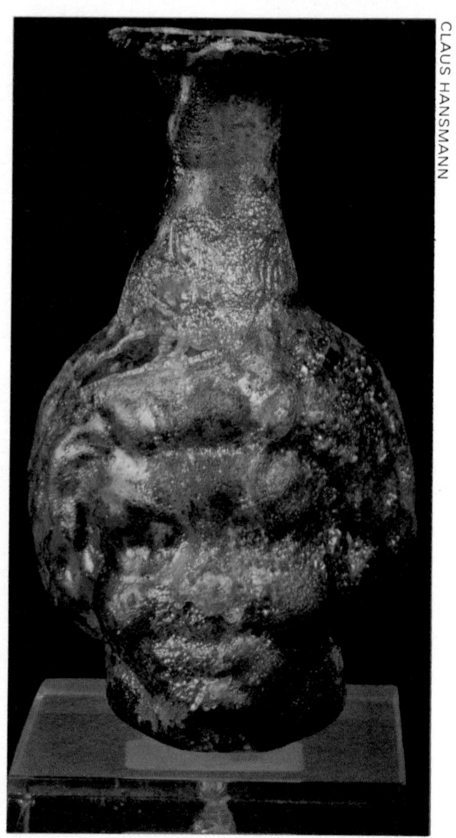

CLAUS HANSMANN

nese ores began to be employed to provide blue and purple glasses, quite apart from the turquoises and greens that copper compounds would give, and that had already been in use for many centuries. The addition of manganese dioxide neutralized the iron oxide, found in impure sand, which gave the glass a dirty brown or yellowish colour.

Decorative effects were produced in many ways. Batches of differently coloured glasses might be mixed to give agate-like materials; or strands, rods and rosettes of glass might be applied to the surface of vessels. Rods of coloured glass might be carefully arranged in bundles and then fused, so that when cut cross-wise they would make decorative designs that could be fused to the surface of glasswares. For exceptional vessels, however, glass cutting, using a lapidary's wheel, was to provide some of the most remarkable glassware ever produced. Thus the so-called Portland Vase in the British Museum was made of a free blown blue glass, the lower half of which was *cased,* or dipped, into a white opaque glass. The latter was then largely cut away with laps to give figures in white relief against the blue of the vessel.

Gatherings of clear glass were blown and opened up while still plastic with tongs opposite the blow-iron. If the blow-iron was then spun rapidly a circular sheet of glass was formed, and from this window panes could be cut, allowing glass windows to be made for the first time.

As in other areas of technology, the Romans made better glass, having obtained the basic technique elsewhere. Roman glass often has fewer bubbles in it than Egyptian, because they apparently were able to fire their furnaces to higher temperatures than previously.

The techniques of glass working, which evolved during the Roman Empire, were kept alive during the early Middle Ages in Byzantium and the Levant, later to be spread again into Europe by way of such centres as Venice, which in the late Middle Ages was renowned for its glasswares.

SCALA

BRITISH MUSEUM/MICHAEL HOLFORD LIBRARY

ANTONELLO PERISSINOTTO

Above: a collection of small Roman flasks, probably used for cosmetics or healing lotions. Glassware of this kind is often found among the remains of Roman buildings and can be surprisingly light and delicate compared with modern glass.

Far left: the Portland Vase, the best known example of Roman cameo glassware. It dates from the 1st century and is named after a former owner, the Duke of Portland. Pottery copies of this vase were made by Josiah Wedgwood in 1790, the origin of the famous Wedgwood style.

Left: a jar made of 'millefiori' glass, of Roman origin, dating from the 1st century BC. This was made by gathering glass canes and cutting them at an angle, giving the 'thousand flowers' appearance. The glass was pressed in a mould to make the jar.

Plate glass

During the Roman Empire, glass blowing, as opposed to moulding or carving, was invented probably near Sidon in the Levant, and widened considerably the uses to which glass could be put. Although known for 300 years, glass had been used mainly for decoration and simple vessels. This new technique opened the way to flat glass and eventually, windows and mirrors. A glassmaker gathered a gob of 'metal' or molten glass on the end of his hollow iron blowpipe and commenced to blow, while at the same time shaping the object either with simple tools or a mould.

At the same time glassmakers gained knowledge of and control over raw materials, allowing first translucent and later colourless glass to be made. And, under the comparatively peaceful conditions of the Roman Empire, the Syrian glassmakers spread across the Mediterranean and north into the forests of Gaul and the Rhineland.

Roman window glass appears to have been cast in a lump and drawn out with pincers into the desired shape while still plastic. Windows were not necessary in the mild Mediterranean climate, but as civilization moved northwards, they became more common. The northern centre of manufacture lay between the Rhine and the Seine; window glass was made by two blowing techniques, the *broad* or spread cylinder and the *crown*. No date is known for the development of these methods: both were known at the beginning of the mediaeval era.

Cylinder, the most widely used process, was based on Lorraine and Hesse. The glassmaker gathered some metal and blew it into a sphere which was then elongated into an irregular cylinder, by swinging to and fro. After reheating, the cylinder was cut lengthwise with shears and flattened into a *sheet* using a block of green wood to smooth the surface.

Crown glass, a speciality of the Normandy glassmakers, apparently developed from the large flat plates made by the Syrians and copied by the early Venetians. The glassmaker blew a sphere as before but attached a *punty* or solid iron rod to it before cracking off the blowing iron. After reheating, the globe would be rotated at speed and the hole left by the blowing iron opened under

Top: plate glass being cast. The molten glass was brought from the furnace in the box-shaped ladle. Once the plate had been cast and rolled with the copper roller, it was slid into the annealing oven to cool down at a steady rate for ten days. The surfaces then had to be polished.

Above: the crucial moment in the making of crown glass. A bubble of glass has been opened out, and by spinning it a disc of fire-polished glass is produced. The central bullseye was used in panes to avoid waste; nowadays fake bullseyes are moulded for use in 'quaint' shops.

centrifugal force into a large flat disc, or *table*. The glass was then cut into quite small panes, including the central crown or 'bullseye', often to be seen on old shop windows.

Crown glass was generally of much better quality than broad: it was thinner, had better surfaces and no flattening was needed, but the panes were small. In the Middle Ages, the great Gothic cathedrals used most of the flat glass produced, originally in self-coloured pieces, though the art of painting on glass dates from the 11th century. From the churches, glass spread first to the houses of the wealthy and then into general use. The panes became larger as did the window openings themselves.

The cylinder process was greatly improved by allowing the glass to anneal—that is, cool slowly before cutting with a diamond and then reheating to flatten it. The improved cylinder was introduced to England in 1832, replacing crown, and was the method used to make the glass for the glass-walled Crystal Palace, London, in 1850.

In 1687 Bernard Perrot of Orleans patented a method for casting *plate* glass which had been developed by Louis Lucas de Nehou. The metal was cast on to a large iron table and spread with a large metal roller. After annealing, the surfaces of the glass were ground and polished to remove roughness, giving a plate of relatively undistorted glass which could be used for mirrors, high-quality glazing, and so on. Cast plate, developed in France, spread slowly to England, Belgium, Germany and, in the late 19th century, the United States.

This development made large pieces of glass feasible. The largest piece of cylinder glass that could be made was about four feet (1.2 m) across, while cast plate glass could be made up to 13 feet (4 m) in length, and half that width.

Through the 19th century window glass continued to be made by the cylinder method although it was gradually mechanized by the use of machines to take the swinging weight and compressed air for blowing. In the 1890s the Americans developed the completely mechanical drawn process which produced cylinders 40 feet (12 m) in height but the cylinder was to be replaced by the modern continuous flat drawn process early in the 20th century. A method for the continuous casting of plate glass was developed in the 1920s, but it has now been superseded by float glass.

RONAN PICTURE LIBRARY

MARY EVANS PICTURE LIBRARY

Top: this English engraving of about 1780 shows the design of furnace used for several centuries. On the left is a sequence of crown glass blowing (reverse order); on the right is a casting table.

Above: making cylinder glass to be used for London's Crystal Palace exhibition hall. The cylinders were slit and opened out while still flexible. The Crystal Palace burned down in 1936.

The crane

R B FLEMING

THE PIERPOINT MORGAN LIBRARY

The crane appears to have been either an ancient Greek or Roman invention for which there is no record before the first century AD. Monuments made of huge stones and constructed before this period, such as the pyramids of Egypt or Stonehenge in Britain, were built without any form of lifting device. As a general rule, when blocks of stone had to be raised an earth ramp was built up which the stones were dragged on rollers, and having been raised to the required height in this way, they were moved into position with simple levers.

Our knowledge of early cranes comes largely from the writings of the Roman architect, Vitruvius, and the engineer, Hero of Alexandria, both of whom were working in the first century AD. The simplest of these cranes was no more than a single pole, one end of which was sunk or fixed in the ground. This beam was raised and held in position by a pair of back-stays attached to its upper end. The pulley-block, which held the hauling line, was fixed to the top. Usually the hauling line was operated by a windlass fixed to one side of the pole near its base, and Vitruvius describes how the same windlass can be used working on the back-stays to raise the beam into position. Rather more complex was a similar crane made of two poles lashed together at their upper ends and with their feet set some distance apart, and thus similar to modern sheerlegs. In such cases a single back-stay was sufficient, while the legs were held in their relative positions by wooden cross-ties like the rungs of a ladder; the windlass could be set between the legs.

Where very heavy loads had to be lifted, the windlass was replaced by a cage treadwheel, similar in design to the treadwheels provided in some cages for mice and hamsters today, in which two

Top: this type of crane, known as the Crow or Corvus, was developed for warfare in the Middle Ages. It had a jib which was raised and lowered by counterbalancing and, when fitted with a 'claw', was used to snatch up the enemy. Alternatively it could be used from outside to lower assault towers inside.

Bottom: Building the Tower of Babel, from Macrejowski's Old Testament, c. 1250. The simple hoist used to haul up the building materials was operated by a treadmill.

or more men could use their weight to raise the burden, while other men could, if necessary, haul on ropes wound round the outside of the wheel.

Hero describes even more substantial cranes employing three and four beams, and he also discusses a form of lobster claw grapple that could be attached to the hauling line. Usually a shallow depression was cut on opposing faces of blocks of stone to accept the points of such grapples, and these pits are still to be seen today in the masonry of many ruined Roman buildings.

The Roman cranes just described had many serious limitations. Although the load could be raised vertically, the degree to which it could be swung to the left or right was severely restricted without pulling the crane off balance. The height to which the load could be raised was limited by the length of beams used in constructing the crane, while the crane was immobile and clearly not very easily lowered and raised again elsewhere.

In time, mediaeval engineers were to overcome the worst of these problems. First, the sheering arrangement was replaced by a vertical wooden post supported by a substantial wooden frame which housed the windlass or treadwheel. The vertical post could be rotated in its framework, and to its upper end were attached a jib and cross-brace. Such a crane could be slewed, since the load, once raised, could be swung to right or left within an arc made by the point of the jib. Furthermore, mediaeval builders realized that such a crane need not necessarily be set upon the ground, but could be raised from time to time as building progressed. To this end a number of gaps were left in the masonry as work went on to accommodate the butt ends of the crane's framework. The crane was, thus, often suspended on the outer face of the wall being built.

Mediaeval cranes, although usually employed for building, were also to become a feature of docks where they were used for the loading of ships.

Left: a large wooden crane at Bruges. Even in mediaeval times it was still worked, as in the Roman period, by a group of men inside an enormous treadmill.

Right: models of a Leonardo crane. Used to winch up stones, it could also be revolved.

Manpower, working a windlass or treadwheel, remained the major source of energy until the advent of steam engines. The German engineer, Kyeser, illustrated a crane worked by windmill sails in the early fifteenth century, an idea probably borrowed from contemporary mills in which the windshaft was used to operate a jib by which sacks could be raised to the top of the millhouse. Kyeser's crane probably never worked, but the German metallurgist, Agricola, was certainly using water power to operate fixed cranes in about 1550.

It is not clear whether Kyeser's crane was intended to be mobile or not, but in about 1480 Leonardo da Vinci illustrated a mobile crane with a counterbalanced jib. The late appearance of this kind of jib is curious since it had long been in use in the Middle Ages. Known as the Crow, it was installed in castles and used to snatch up besiegers if they ventured too close. In 1139 Prince Henry of Scotland narrowly missed being caught up in this way. Equally, the besiegers occasionally used similar cranes to lower small assault towers over the walls of fortifications under attack.

SCALA

Balances and weights

Few balances have survived from periods before the Roman domination of Europe, although their beginnings are estimated to date back to 5000 BC. The first evidence of their existence comes from stone or copper weights which were found on archaeological sites. These weights, often in the shape of animals or birds, began to be used in Mesopotamia and Egypt shortly after 3000 BC, their values being multiples of a common unit, the weight of a grain of wheat. This fact alone suggests that even before stone and metal weights came into use, wheat itself may have been used as a measure of weight. This in turn implies that early balances were evolved for measuring small quantities of precious materials, such as gold or silver. The earliest illustrations of balances, from about 2000 BC, support this theory because they invariably show balances being used for weighing precious materials.

Early balances were of simple construction. They were little more than a beam pivoted in the middle with a pan suspended by cords at each end. Early illustrations give very little information about the finer details of construction. In Minoan Crete of about 2000 BC for instance the balance was used as a word-sign meaning 'weight' or 'to weigh' and the symbol itself was drawn with a minimum number of hasty lines. Since the accuracy of a balance depends largely upon the nature of the pivots, there is no direct evidence of how precise these instruments were. Even so, with the grain of wheat as the basic unit of weight, a fair degree of accuracy must have been obtained.

The political conditions of the Near East at this time were such that each city state developed its own system of weights. For this reason denominations of weight larger than the grain varied from place to place. The shekel, for example, might have been reckoned at anywhere between a hundred and two hundred grains depending upon the local ruling. A merchant, therefore, travelling in the eastern Mediterranean would have to carry a set of weights

SCALA

Above: a typical polished stone weight from Ur, Mesopatamia, dating from about 2500 BC. This design, of a duck preening itself, was common at that time, although the symbolism involved is a mystery. The value of the weight is carved on the side.

Below: Egyptian wall paintings—this is a copy of one at Thebes—often showed balances like this, but usually failed to make interesting details of the suspension and adjustment clear. In this case tubular ingots of gold, in the left hand pan, are being balanced by a weight in the shape of an animal's head. Egyptian weights were often simple blocks, however.

PHOTO: JOHN FREEMAN

appropriate to each port of call, and sets of weights of this kind have been recovered recently from the wreck of a merchant ship off the southern coast of Turkey. People engaged in this sort of trade must have been capable of complicated mental arithmetic which allowed them to convert from one of the major systems of weight denomination to another.

In one Egyptian wall painting, made shortly after 2000 BC, there is a hint that a major innovation had been introduced into the design of the balance. In this picture a man is shown moving or placing an object on one arm of the balance beam. This could be either a small weight used to adjust the uneven balance of the instrument, or it could be a tare (a kind of plant from the pea family) used to offset the weight of a basket or other container in the pan on the opposite side of the balance. It seems more likely, however, to be a sliding weight, or *rider*, used to measure the smaller denominations, in which case the beam would have been graduated with a scale.

The use of riders was to lead to the development of a balance that became very common in the Roman world, the steelyard balance. This instrument was designed in such a way that the weighing pan on one side of the support pivot exactly balanced a long, graduated beam with a rider set at zero. Increasing weights were then measured by sliding the rider outwards. A considerable number of such balances, usually made of bronze, have survived, and in many cases a knife edge fulcrum (pivot) was employed. This refinement reduced friction to a minimum and provided greater accuracy. A hole was made in the beam at the point of pivot and the fulcrum, in the form of a short knife blade, was passed through it.

Greek science, which developed largely from 500 BC onwards, demanded precise instruments. This was especially true after it had been established that gold could be assayed (checked for purity) by measuring its specific gravity. This was done by accurate weighing of an object first in the air and then suspended in water.

From the eighth century AD Arab scientists, working at first from translations of Greek writers, began to improve the design of balances. After this time, for example, manuscripts dealing with the description of precious stones often included surprisingly accurate estimates of their specific gravities, while a small number of treatises in Arabic dealt exclusively with the design of various types of balances to be used for scientific purposes. Translations of these books in Arabic made in the 12th century gave European scholars the opportunity to learn to construct accurate balances, although simple balances had been in use throughout the early part of the Middle Ages.

CLAUS HANSMANN

Above: a Greek weight with a design which shows the god Mercury, cornucopias or horns of plenty and an ear of grain. The inscription reads 'public weight of the year four'. It weighs 25 oz (690 g).

Below left: a German corn balance of the seventeenth century, made of brass. The weight was measured by moving the cube-shaped rider along the graduated arm. This steelyard type of balance was made illegal in England beacuse an unscrupulous trader could alter it to give false readings.

Below: a replica of a Viking balance, found in a hoard on the Baltic island of Gotland. This balance, made of bronze, was of a folding portable variety, used for weighing gold and silver. The stand is a modern addition—the balance would have been suspended by hand.

CLAUS HANSMANN

PHOTO: JOHN FREEMAN

The loom

MANSELL COLLECTION

Above: a Greek vase from about 450 BC, showing Telemachus, Penelope and a vertical loom. Weights keep the threads taut.

Right: 'The Romance of Alexander', a fictional work written by Eustache of Kent about 1250, shows this heddle loom.

Bottom: Chinese silk weaving in the nineteenth century. The person sitting behind the loom would raise the heddles.

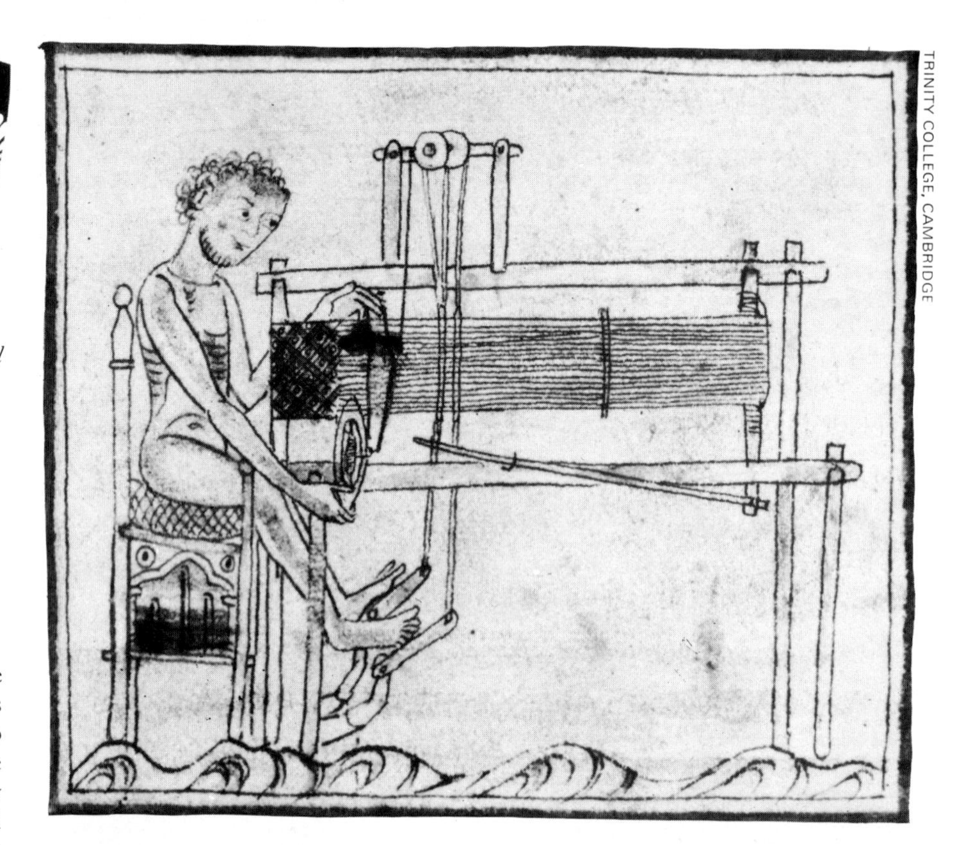

TRINITY COLLEGE, CAMBRIDGE

GIRAUDON

A loom is any device in which one group of parallel threads, the *warp*, is held under tension while a second group of threads, the *weft*, is woven into the warp. In its simplest form a loom may be little more than a pair of parallel horizontal beams set some distance apart and supported by a peg at each end which is driven into the ground, the warp threads being stretched from one beam to the other. Looms of this type, ground looms, are still used by many primitive races today, while the earliest picture of such a loom is to be seen on an Egyptian dish of about 4000 BC. Only slightly later is an engraving of a similar loom on a seal-stamp from Mesopotamia.

The art of weaving, and therefore the history of the loom, is however far older than these early pictures. Small fragments of woven cloth have been found on many archaeological sites from early in the New Stone Age, from 8000 BC onwards, while the stone or bone *whorls* (flywheels) used on spindles for spinning the thread are often found on sites of the same age.

A major problem that early man had to solve was how to weave in the weft threads rapidly, other than by passing them under and over each warp thread in turn. A bar, known today as the *shed rod*, which ran from side to side passing under every second warp thread, was used; by raising the rod, a space (the shed) was formed between

Below: a Navajo Indian rug weaver. At one time, the loom would have been strung between two trees. The flat wooden rod is the weft bobbin.

Bottom left: Turkish girls from Cappadocia weaving a carpet on a loom very similar to the Indian one above.

Bottom right: a nomad Turkish woman weaving a kilim (tapestry) among Roman tombs to the west of Tarsus.

SUSAN GRIGGS/ADAM WOOLFITT

the warp threads through which the weft bobbin was passed. The return, however, still had to be woven in and out by hand, but this difficulty was eventually overcome by attaching those threads lying below the shed rod to a second bar, the *heddle*, by a series of loops of cord. By raising the heddle a new space was formed between the warp threads, which allowed the weft to be returned. The heddle must have been invented before 4000 BC since the Egyptian painting already mentioned shows three heddles lying across the warp.

The ground loom had one major disadvantage in the amount of space it required. Thus in Egypt and other countries of the Near East it could be pegged out in the open, but in regions with less reliable weather another type of loom was evolved that could be used indoors. The ground loom was set upright against a wall and supported by a pair of vertical posts. The warp threads were suspended from a beam at the top of the posts, but the lower beam was usually replaced by a row of clay or stone weights as these kept the warp threads under more even tension. Looms of this kind were first known in Europe in the early Bronze Age around 2000 BC and continued to be used until about 1200 AD.

The Chinese began to cultivate and weave silk in about 500 BC, and in order to produce a cloth with complex patterns woven into it, they had to use a loom with many heddles. Stopping repeatedly to raise different heddles must have proved tedious, for by 200 BC the Chinese had designed a loom in which the heddles could be raised by a system of cords and pulley wheels operated by pedals, thus leaving the weaver both hands free for his work. In this loom, too, the warp threads were wound on to one beam and could be wound off as the work went on, the finished cloth at the same time being rolled on to the second beam. It was this loom in a modified form that spread first to the Arabic speaking countries, and then to 13th century Europe.

Many people in antiquity were nomadic herdsmen, tent dwellers constantly on the move, and for them heavy looms would have been impossible to pack and transport. The nomads of Arabia and Central Asia thus evolved a loom which had one end attached to a tent post, the warp threads being kept tight by a belt around the weaver's body. Heddles were replaced by a pack of bone or wooden tablets which had holes in the top and bottom for threading the warps through. By twisting the pack backwards and forwards warp threads could be raised and lowered to allow the passage of the bobbin. Cloth made on these looms was always narrow, and strips had to be sewn edge to edge if a broader cloth was required. Pieces of cloth and tablets have been found in Central Asia dating from about 1000 BC. This type of loom was later adopted by more settled people in Europe and the Near East.

SONIA HALLIDAY

SONIA HALLIDAY

Bricks

Twelve thousand years ago men of the Old Stone Age were still nomadic hunters, but they occasionally built themselves crude shelters of wood and stone. The floors and walls, and sometimes the fireplaces, were daubed with sandy clay and burnished to make them smooth. In the New Stone Age which followed, when men in the Middle East began to become settled farmers, more substantial structures were needed. Builders had to turn their attention to existing materials, and in many parts of the Middle East neither suitable stones nor timber were available.

In the river valleys, however, there was plenty of mud, which was used to make the first bricks. In some places such as upland Persia and Palestine the earliest bricks were simply fashioned by hand into oval lumps. Some, for example those from Jericho, were made with a series of notches on the upper surface to provide a keying for the mud mortar used when building. Such bricks were not fired but sun-dried until hard.

At Catal Hüyük in Central Turkey, one of the earliest New Stone Age villages so far discovered, bricks were shaped by quite a different method. The wet mud was rammed into a wooden frame which was then raised, leaving a rectangular brick lying on the ground where it could dry in the sun. Gradually this technique was adopted throughout the whole of the Middle East, and an Egyptian tomb painting of about 2000 BC depicts the entire process from mixing the clay to stacking the dried bricks.

Usually a sandy mud was chosen for brickmaking as this reduced the hazard of the bricks cracking when they dried. In some places, however, the clay was too fine and sand or straw had to be added to make a usable brick, hence the lament of the Israelites in Ancient Egypt that they could not make bricks without straw.

Sun-dried bricks provided a cheap and readily available building material and entire cities, including the defensive walls, were constructed of it. It had, however, one major disadvantage: it disintegrated in heavy rain, and so required constant repairs to buildings. For this reason temples were often made of stone, despite the high cost of transporting the material. In about 2500 BC in Mesopotamia a new solution was found when builders began to fire their bricks. Fuel, however, was scarce and often fired bricks were used only for the quoins (cornerstones), jambs and arches,

SCALA

Left: brick sculpture dates from about 1590 BC in Babylon. The bricks were usually moulded in wooden or stone moulds and coloured with glazes. The figures shown here are Persian archers, from a frieze depicting the Royal Guard of the Persian Kings at the Palace of Darius, and dates from about 500 BC.

BRITISH MUSEUM/MICHAEL HOLFORD

Below: a model of brickmakers in Egypt, found at Beni Hassan. The man in the centre holds a tool which was used for digging out the clay. After puddling with water the wet clay was put into moulds.

RONAN PICTURE LIBRARY

Above: general view of a brickfield. In the foreground bricks are being hand moulded and placed on a barrow to be taken to be placed in piles (hacks) for preliminary drying. A horse-powered pug mill grinds the clay.

or a facade of fired brick might be applied to an otherwise mud-brick building.

The idea of firing bricks to make them harder and more durable was a logical development, for men had already been using pottery for several thousand years beforehand. The bricks were set in a pile on the ground, fuel was built up as a *clamp* (a pile of fuel) over them and lit, a method also previously used for firing rough pottery.

The peak of brick manufacture in the Middle East was undoubtedly reached in Assyria and Babylon in about 500 BC. Facing bricks were now not only fired but given a coloured glazed surface, the glaze itself being the same material as that used for glassmaking in both Egyptian and Mesopotamia from about 2000 BC onwards. Figures of animals and warriors were sometimes modelled in low relief, many bricks being required to form each figure.

ZEFA

Top right: sun-dried bricks made from mud and straw were one of the earliest domestic building materials and are still used today by the people of Niger in the Sahara desert. The straw helps to strengthen the mud.

ZEFA

Bottom right: after the wooden mould is removed from the wet clay, it is dried to leather hardness before firing.

RONAN PICTURE LIBRARY

Above: hand moulding of bricks in the early 19th century, a skilled job. The moulder throws a lump of clay, previously rolled in sand, into the mould so that it fills all the corners.

Pumps

One of the major needs of the ancient civilizations of Egypt and Mesopotamia was to irrigate their crops during the growing season, and this meant raising water from the rivers into channels to supply the fields along the valleys. At first the water was simply bucketed out, an inefficient and back-breaking use of manpower. Shortly after 3000 BC, however, the people of Mesopotamia found a way of raising water that took much of the tedium out of the work with the invention of the *shaduf* or swipe. A long wooden lever was pivoted on a pair of upright posts set close to the river bank. A slender pole with a bucket suspended from it was attached to one end of the lever, with a counterweight fixed to the other end. The operator had only to push the pole downwards to fill the bucket, which would be raised largely by the counterweight and then emptied into a trough leading to the irrigation channel. This simple device spread throughout the Middle East and is still to be seen in use today. From the time of its invention it remained the only water-lifting machine for over two thousand years.

Shortly after 500 BC three new water-lifting machines had been invented, but unfortunately very little is known about how they came to be created, or even where they were originally used.

The first of these inventions, the water-raising wheel or *saqiya,* was in its simplest form a large wheel with a series of pots lashed to its circumference. The wheel was mounted so that the lowest pots dipped into the water and were filled. As the wheel turned these pots were raised, and at the top emptied themselves into a chute leading to the irrigation channel. A natural development of this principle was to make a wheel entirely of wood, the pots of the original being replaced by troughs. A water-raising wheel of this kind is described by Vitruvius in the first century BC. The same writer mentions yet a further development: in the so-

KEYSTONE PRESS AGENCY LTD

RONAN PICTURE LIBRARY

Top: the design of the shaduf, seen here in Egypt, has changed little in nearly five thousand years.

Right: the principle of the lever as a labour saving device was still used in this balance force pump of 1816.

called *tympanum* the wheel itself is made of a series of radiating watertight compartments, each supplied with an aperture for filling and discharging.

The second machine was a continuous loop of chain passing over a pulley-wheel, with buckets attached at intervals. It operated in the same manner as the saqiya, and some believe it was with the bucket chain that the Hanging Gardens of Babylon were watered (600 BC). There is, however, no direct evidence of its existence until the writing of Vitruvius.

The third machine used the Archimedean screw or *cochlea,* so named because it resembled a snail. This was made by cutting a deep screw-thread into a heavy cylinder of wood which was then encased in a water-tight coating of planks. A bearing on one end of the cylinder was passed through an upright post just above the level of the river. The other end was similarly mounted on a higher post on the river bank, and made to turn by pedals. As it rotated water was carried up the thread of the screw and discharged into a chute at the top. Although Archimedes of Syracuse, who died in 212 BC, is said to have invented this pump, it is more commonly believed that he saw it in operation in Egypt and through his teaching popularized it. By the time of Vitruvius it was commonplace throughout the Roman Empire.

The ancestor of most modern pumps, however, was a force pump incorporating a cylinder, plunger and valves, designed according to tradition by Ctesibus of Alexandria early in the first century BC. This type of pump was made of bronze and was, again, described by Vitruvius. It is beyond question that such pumps were made and used because fragments have been recovered from a number of Roman buildings, including an almost complete pump from Bolsena in Italy.

The important innovations in this kind of pump were the valves and the plungers. Valves already existed in the wooden bellows used by blacksmiths, while many other machines invented by the technicians at Alexandria incorporated plungers. The pump had to be precisely built of bronze and this proved too expensive for most everyday uses. It was, however, employed as a weapon. When the Arabs attacked the late Roman Empire the Byzantines used force pumps to hurl inflammable liquids—'Greek fire'—at the enemy.

KEYSTONE PRESS AGENCY LTD

Above: a reconstruction of the Bolsena force pump. There were valves at the base of each upright cylinder and in the pipes on either side of the central outflow.

Below: a Chinese solution to the problem of raising water for irrigation was this chain pump, a design which was probably invented in the 2nd century BC.

MARY EVANS PICTURE LIBRARY

Smelting and alloying

Long before man began smelting metals from their ores he had been using the bright blue and green copper minerals as paint pigments. During his search for these colours he had come across small pieces of metallic copper, so-called 'native' copper, embedded in the ores. This copper was used to make trinkets in various places in the Near East early in the New Stone Age, from about 5000 BC, onwards. They were shaped by the only means then known, by hammering and bending. About this period small pieces of gold began to be worked in the same way. Later it was discovered that copper became brittle as a result of hammering, but it could be softened again by making it red hot.

How, when, and where it was discovered that copper could be extracted from its ores is unknown. Because the process demanded fairly high temperatures, it can be assumed that smelting evolved as a by-product of pottery production. The process as it is first known involved heating a mixture of charcoal and copper ores in a small hollow in the ground. The temperature was raised by a number of men with blowpipes puffing into the fire. This system is clearly depicted in a number of paintings found in Egyptian tombs dating from about 2500 BC onwards.

In this simple process the molten copper collected in the bottom of the primitive furnace and, after cooling, could be recovered as an ingot shaped like a bun. With the discovery of smelting came the realization that copper liquefied on heating, and could then be poured into prepared moulds. This meant that the 'bun' ingots could be broken up, placed in crucibles and reheated for making castings.

This basic process was to change little over the next thousand years. The capacity of the furnace, though, gradually increased by building up a chimney-shaped structure with stone and clay, and several simple types of bellows evolved to replace lung power to fan the flames. These bellows were often no more than a skin bag with a hand operated valve or a drum with a slack skin covering that could be raised and depressed to create a draught. It was not until about 500 BC that bellows with hinged valves came into use.

Shortly after the discovery of copper smelting and casting, silver began to be extracted from its ores. At the same time gold was being melted and cast. Possibly as a result of mixing some of these metals in a crucible, it was discovered that mixtures of metals (alloys) are usually harder than some of the ingredients that go into their making. From about 2500 BC onwards various alloys of silver and copper or gold and silver were used for jewellery throughout Mesopotamia and Egypt.

Although plentiful, the chief defect of copper as a material from which to make tools was its softness. In theory it could have been made harder by additions of gold or silver, as was done in pre-Columbian Peru, but this would have been costly. Instead other, less expensive metals were sought for alloying with copper. At first tin, antimony and arsenic, which form ores that are heavy and similar in appearance, were used for alloying with copper. By 3000 BC, however, tin, the commonest, became the one generally used to give the alloy bronze.

At the same time that bronze began to be produced, lead was also extracted from its ores. It was such a soft and dull metal that it had few uses at first. By about 1500 BC, however, it was discovered that by mixing lead with bronze the alloy could be made to pour more easily when casting. From this time onwards larger and more complex castings were possible. This alloy of copper, tin and lead is still used for casting statues.

It was the Romans, however, who used large quantities of lead to line their aqueducts and to pipe water. It was the Romans, too, who introduced the alloy of lead and tin (pewter), used to manufacture dishes and cups as a cheap substitute for silver. The same alloy was also used as a solder for joining pieces of bronze or copper.

Tin was never plentiful, and possibly because of this, Persian craftsmen in about 500 BC began to use zinc in

This wall painting in a tomb at Thebes, dating from about 1500 BC, shows the casting of a bronze door. At left, foot bellows are being used, while at the centre the molten metal is being poured into the mould through small funnels.

ERICH LESSING-MAGNUM PHOTOS

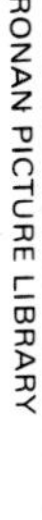
RONAN PICTURE LIBRARY

STAATSBIBLIOTHEK BERLIN

place of tin, thereby creating the alloy known today as brass. The large scale production of brass, however, began after the Romans had conquered Europe, and by 100 AD they were exploiting the Central European deposits of zinc ore (*calamine*) so that many of the later Roman coins and ornaments were made of brass rather than bronze. As a result brass became the common copper alloy of the Middle Ages, although the method used to make it was different from today's technique. Zinc vaporizes very easily in a furnace, and so to make brass, the zinc ore was thrown into molten copper and stirred to form the alloy.

Iron was also difficult to produce because its melting point was too high for the temperature produced by the simple furnaces. The metal remained in the furnace as a spongy mass (the *bloom*) and this later had to be removed, heated and hammered hard to form a bar of metal which could only be shaped by further heating and hammering. Small quantities of iron were made by 2000 BC, but it took another thousand years before its manufacture was completely mastered. From then on, because its ores are very common, iron became the normal metal from which to make weapons and tools, so replacing bronze for such uses.

Because iron was not molten at any stage, it could not be alloyed in the way that other metals were. To overcome this, iron bars were heated in charcoal over a long period and the bars then absorbed some carbon to give a form of steel, a fact discovered early in the Roman period. When Caesar invaded Britain he was amazed to find the local tribes still using iron swords which became bent in battle. His own troops were using steel weapons which were harder and less flexible.

Forced draught provides a good supply of oxygen, needed to reach high temperatures. An early method was to use bellows made from animal skin, with the draught blown through an open leg. A pair of goat skin bellows are seen in the top picture, taken in Nigeria in 1964 in the workshop of a sword and razor maker. The lower picture is of a more advanced furnace, shown in a book of 1697. This shows a bronze gun foundry, with a furnace in which bellows blow the flames across the roof; heat from the roof melts the metal below, taking over a day to do so since the furnace could hold up to 30 tons at one loading.

Casting of metals

Man first began to take an interest in naturally occurring copper and gold about 6000 BC at the end of the New Stone Age. The only method he had then for shaping these metals was to hammer them. Early illustrations of Egyptian goldsmiths at work, dating from about 2000 BC, show how the metal was beaten out on a stone anvil using a rounded pebble as the hammer, which was held in the hand as it had no *haft* (handle). As a system of shaping, hammer-work was excellent for producing such items as bowls or gold leaf. To this day hammer-work is still used. The modern silversmith does much of his shaping by hammering, while in the final stages of gold leaf manufacture the thinning down is still done by hand hammering.

Shortly after 5000 BC it was discovered that pure copper and silver could be extracted from their ores by heating. This led to the discovery that, if heated sufficiently, metals would melt, and as liquids they could then be poured into moulds where they would set on cooling. This process offered immediate advantages over hammering because it made it relatively easy to form solid objects such as axes and daggers. Furthermore, if the mould itself were made to carry decoration, so too would the object cast in that mould. This discovery greatly simplified the process of making gold and silver jewellery.

The earliest known moulds were made of stone. For this purpose soapstone was very popular, for not only was it soft and easy to cut, but it also stood up well to the sudden heating during casting. In the European Bronze Age, from about 2000 BC onwards, soapstone became so popular that it was exported in considerable quantities from Norway to Denmark and northern

Right: copper ingots recovered from the earliest known shipwreck, a trading ship that sank off the coast of southern Turkey in about 1200 BC.

Below left: four stages involved in making a cast bronze sword. From left to right: the wooden prototype used to shape the mould; the mould itself; the bronze sword in the mould; the finished sword.

Below right: a Chinese bronze casting.

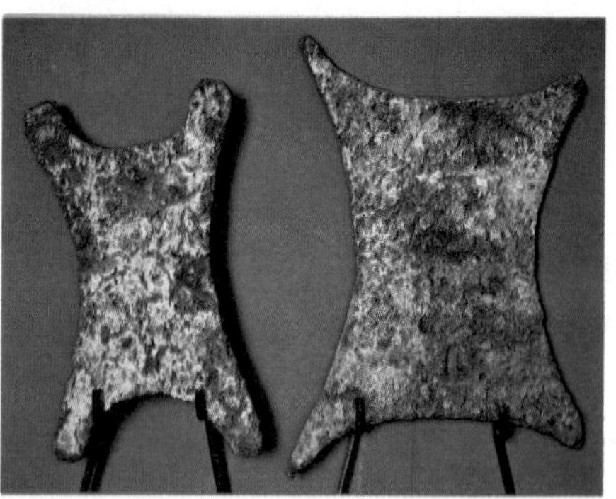

SONIA HALLIDAY

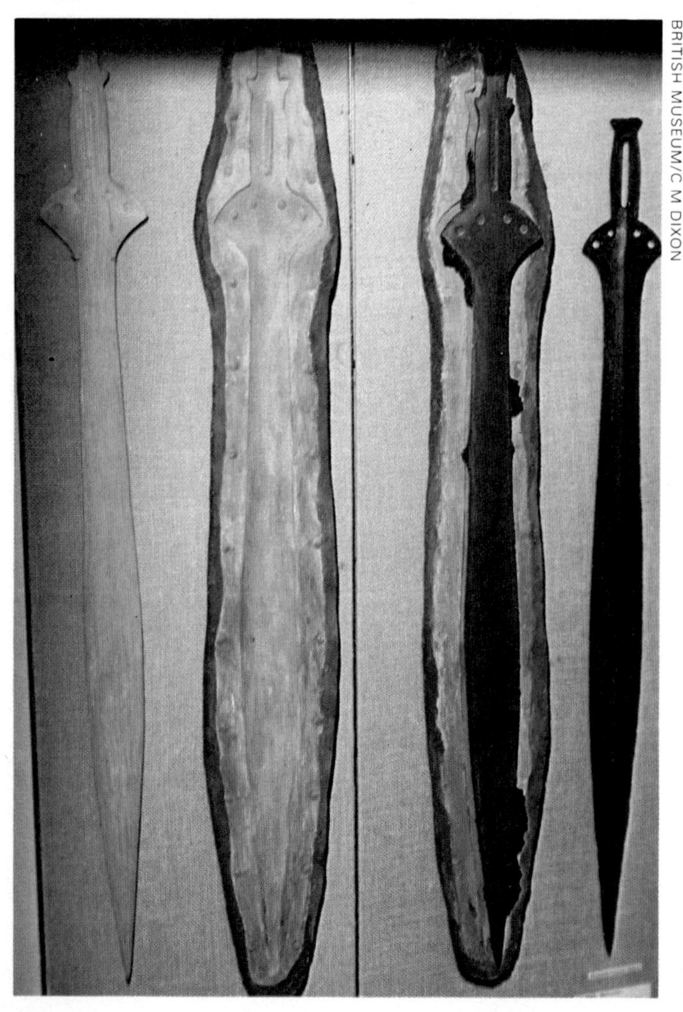

BRITISH MUSEUM/C M DIXON

BRITISH MUSEUM/MICHAEL HOLFORD

Germany just for making moulds. A second material used for moulds dating from about 4000 BC, was sandy clay, which overcame the need for cutting, since the clay was plastic and could be wrapped around a model of the object to be made. After drying, the clay was fired as though it were pottery and was then ready for use. Both stone and clay moulds tended to wear away with repeated use, and where many objects of the same shape, such as arrowheads, were required, bronze moulds were used. Curiously, one of the earliest uses of cast iron in China, in about 500 BC, was for making moulds in which bronze axe heads were cast.

The design of moulds changed considerably between 5000 BC and 3000 BC. At first the *matrix* (the mould) was simply a hollow in the shape of the object to be cast cut into the face of the stone. This was sufficient as long as the objects were flat or of a simple shape, but more complex objects, though, required two pieces of stone or clay made to fit face to face, with the matrix hollowed out of each piece. In practice the two pieces known as piece-moulds never fitted perfectly and a thin web of metal squeezed its way between them during casting. The remains of these webs can be seen on many early cast tools and weapons, just as they can be seen today on plastic toy soldiers which are also cast in piece-moulds.

Some bronze implements such as axes and spears needed a socket for the handle. This was done by making a cone of clay that was held between the two pieces of the mould. After the casting had been made this clay core could be dug out to provide a socket. By 200 BC the Chinese had become adept at making clay piece-moulds and cores, using not just two pieces but often as many as 30 pieces of clay fitted together. Cauldrons and other bronze vessels made in China at this time are masterpieces of casting.

In the Middle East from about 2000 BC onwards another method of making moulds known as 'lost wax' casting began to be used. First a wax model of the object to be cast was made, or a wax model might be built up over a core of clay. The model was then coated with layers of sandy clay and, when dry, was fired as if it was a piece of pottery. This made the clay brick hard and in the process the wax at first poured out and finally burnt away. The molten metal was then poured into the hollow left by the wax.

This lost wax casting had the advantage that it could produce very elaborate castings, and it is still used by art foundries today. Its only drawback was that since the mould had to be broken open to get at the casting it could not be used a second time. Further copies could only be made by creating a fresh model in wax on each occasion. The process was therefore not used for producing weapons or tools, but for casting ornaments and statues.

Right: the cat was worshipped by the ancient Egyptians, and large numbers of cast bronze statues of cats were made, usually by the lost wax method. The one shown here was made in about 600 BC.

Below left: the Etruscans were the most powerful race in Italy in the second quarter of the first millenium BC. Where they came from originally is not known for sure, but one theory is that they were migrants from Asia Minor. They were skilled artists and craftsmen who took a great interest in natural science. This bronze casting of a wolf and cub was made between 700 and 600 BC.

Below right: the two halves of a mould used for making cast bronze axe heads.

BRITISH MUSEUM/C M DIXON

BRITISH MUSEUM/MICHAEL HOLFORD

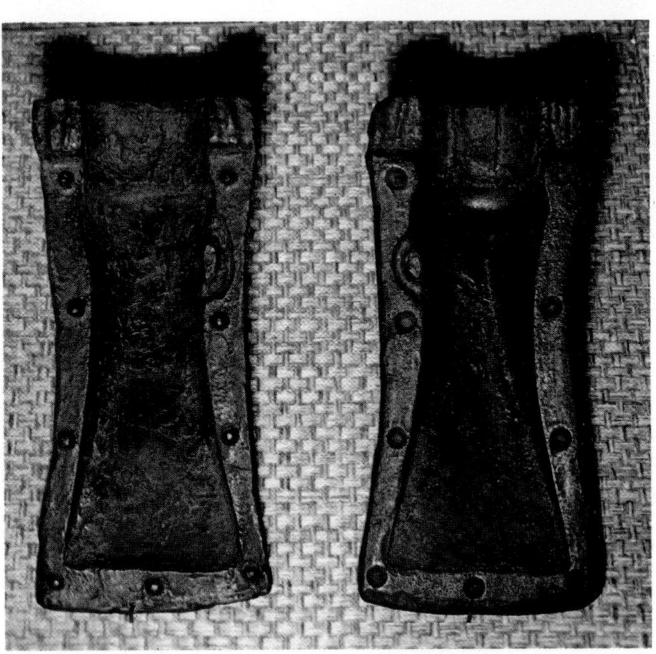
RIJKSMUSEUM/C M DIXON

Coinage

Among primitive people today 'money' takes many different forms. In some parts of the world it represents real wealth such as cattle or bars of metal. Thus in East Africa, until quite recently, pieces of iron shaped like the blade of a hoe were regularly used in transactions. In other parts of the world money may have little intrinsic value, as for example the cowrie shells used in the South Pacific Islands. Both kinds of money, however, share one thing in common: the people who use it accept it as a means of exchange in the buying and selling of goods.

In the countries of prehistoric Europe and the Middle East there existed similar forms of money. Thus from about 2000 BC onwards one finds copper ingots, usually in the form of a miniature ox hide, being traded widely. Bronze was often made up in the shape of neck-rings, while after about 700 BC long, narrow, iron bars began to be used as currency. Curiously, gold and silver do not appear to have been used in these early periods as money, although the reason for this may well have been the state of metal technology at that time.

Gold and silver are easily adulterated

MICHAEL HOLFORD LIBRARY

SNARK INTERNATIONAL

A selection of silver coins from around the 5th century BC. Early coins often bore designs depicting a city's staple product or sacred animal on one side and the head of a deity on the other. Top: a Greek coin from Naxos, Sicily, famous for its grapes. Bottom right: an Athenian tetradrachm coin bearing the owl of Athens on one side and the head of the goddess Athena on the other. Bottom: an early Hebrew coin, probably of Greek origin but bearing a Hebrew inscription.

with copper, and it was not until a method of testing gold with *touchstones* and of purifying the metals by *cupellation* had been evolved that it became a practical possibility to create gold and silver money. Touchstones are black stones related to flint, which were used to test gold and silver for purity by the mark left on the stone after rubbing the metal. Cupellation is a method of refining precious metals by melting them and blasting them with air to oxidize or blow away impurities.

According to legend, the first coins were produced in Lydia in western Asia Minor in about 700 BC, and indeed a small number of gold coins of this period exist to this day.

The method first adopted to manufacture coins is of interest since it is basically the means still used. First, the gold was weighed out to a precise quantity. Each weighing would be made up of a number of small fragments of the metal. The weighings were then placed in shallow, circular depressions in a clay mould, and were heated until they fused and ran to cover the base of the depression. On cooling these buttons or blanks of gold were removed, placed on an anvil and stamped (or struck) with an iron die bearing the symbol of the state or king who was issuing the coins. The symbol might be a profile of the king's head, or it might have some religious significance such as the owl that appeared on early Athenian coins. Often a pair of dies was used, one on each side of the blank, so that when struck an impression was made on each side of the coin (*obverse* and *reverse*).

The reason for stamping these early coins was to provide a form of guarantee as to their weight and purity. Even so, the earliest gold and silver coins do not appear to have been used as currency, but rather as a convenient form in which to handle bullion. This is not to say that these coins were never used in business transactions, merely that even after their introduction trade was normally conducted by barter and exchange.

A true money economy in which it became accepted that goods could be exchanged for cash appears to have developed in the second century BC, and its growth was clearly accelerated by the expansion of the Roman Empire. It was now no longer possible to mint only gold and silver coins, and smaller denominations were struck in bronze. The method of production, however, remained the same.

SCALA

Above: the mountain nymph Arethusa is shown here. A signed coin from Syracuse.

Below: Indian gold coins. Left, Gupta coin c.400 AD. Right, Cholas coin c.1000 AD.

The remains of local mints have been discovered during the excavation of many Roman cities, and it is clear that the issuing of coin was under the control of local governors. This practice could clearly lead to abuses.

In the first century AD, however, the composition of bronze coinage was changed by reducing the quantity of tin in the alloy and replacing it by an even larger proportion of zinc. When new these coins would, like the Pinchbeck alloy (an alloy of copper and zinc used to make imitation gold jewellery named after the eighteenth century English watchmaker who invented it), have appeared almost golden, despite the fact that they were cheaper to produce than a tin bronze.

The making of counterfeit coins is almost as old as the minting of money. The forger, however, had only two alternatives open to him. Either he could follow the original process and cut counterfeit dies with which to strike the blanks, a very difficult thing to do, or he could prepare a mould from a genuine coin and cast his metal into it. In this case the forger would have to remove the flashes formed on the edges of the coin at the junction of the two halves of the mould. In either case a careful scrutiny of the coin could show it to be forged, and it is for this reason more than any other that the original process of striking blanks was retained.

Iron and steel

All the iron made in Western Asia and Europe from 1000 BC onwards was originally made by the bloomery process. The iron ore was heated in a furnace to a temperature of about 1100 °C (2010 °F), sufficient to remove any silica in the ore as liquid slag, but leaving the iron in the furnace as an unmolten spongy mass, the *bloom*. The bloom was later removed, reheated and forged to consolidate it into a workable piece of wrought iron. Such metal was soft and easily bent, and Caesar, commenting upon the weapons of the Britons, noted how they had to retire from the battle from time to time to straighten out their swords.

Undoubtedly the Romans understood that by putting wrought iron in a forge in contact with charcoal for a long period it would gradually become harder and less flexible. Today this process is known as case-hardening; brought about because the iron has absorbed a limited amount of carbon to form a component of steel, cementite. The process was, however, very slow, and the layer of cementite formed on the surface very thin. By repeatedly case-hardening a bar of iron, cutting it into short lengths, piling the pieces on top of one another, hot forging them to weld them together, drawing the bar so formed down thinly, and again case-hardening it, a bar of laminated metal was formed in which layers of iron alternated with layers of cementite. This material had some of the properties of both of its components, being sufficiently hard and flexible for making weapons. Indeed, the traditional Japanese Samurai sword was made in this way, with often as many as a hundred successive piling and cementation processes being carried out to produce the fine laminated structure. In Saxon and Viking Europe several bars of laminated metal were often twisted and welded together to form the core of swords which, when polished, showed a characteristic herring-bone pattern down the length of the blade. Early Islamic travellers who encountered the Vikings were greatly impressed by the quality of these swords.

A totally different approach to the manufacture of steel was pioneered in southern India. By 500 AD the Indians were making crucible steel, often referred to as *wootz*. Bloomery iron was broken up into small pieces and packed into crucibles with charcoal. The crucibles were covered with lids which were *luted* into position with clay to make them airtight. A stack of such crucibles was then placed in a furnace and kept at near white heat for as long as a week. During this heating the iron absorbed sufficient carbon to produce the component of steel that is called pearlite today. After heating, the fragments of metal were hot forged to form a bar of steel that when polished had a typical watered pattern on the surface.

In the late Roman period *wootz* steel was exported from southern India and brought by Arab seamen and caravans to the eastern Mediterranean. The Roman emperor, Diocletian, set up an arms factory in Damascus partly to shape the imported *wootz*, which the Romans believed to come from China. The products of this factory were referred to as damascened steel, and after the Arab conquests of the eighth century other factories were set up, one of them at Toledo, Spain.

BURY PEERLESS

WERNER FORMAN ARCHIVE

Above: a steel glove with iron mail, decorated with silver and gilt, which was made in Jaipur, India, in the 18th century.

Left: the 4th century iron pillar which stands about 8 miles (13 km) south of Delhi, India. It is about 24 ft (7.3 m) high, tapering from 16.4 in (42 cm) diameter at the base to 12.5 in (32 cm) diameter at the top, and shows no signs of rust despite its great age.

JOHN FREEMAN

TRANSWORLD FEATURE SYNDICATE

In China the development of steel manufacture followed yet another pattern. By 500 BC the Chinese had learnt how to produce cast iron. This was due largely to two factors. Many deposits of iron ore in northern China contain high proportions of phosphorus, the presence of which reduces the temperature at which iron melts, although the resulting metal is unduly brittle. Equally, plentiful outcrops of anthracite coal allowed the Chinese to reach furnace temperatures of 1200 °C (2192 °F) and above quite easily. Hence, the early Chinese iron founders treated the metal much as though it were copper or bronze, and cast it in moulds to shape it. This brittle metal, although adequate for the making of vessels and agricultural tools, was useless for the manufacture of weapons. By 200 BC, however, the Chinese had learned, probably from their northern neighbours, the Mongols, how to make wrought iron, and shortly after this date they began to make steel by a process of co-fusion. Wrought iron and cast iron were broken down into small fragments and heated in a reverbatory furnace to form a pasty mass. During this heating the cast iron component lost much of the carbon it contained, while at the same time the wrought iron absorbed some of the carbon. If the process was correctly carried out the end product should have been a good quality steel composed largely of cementite and pearlite. As it happened, even as late as the tenth century AD, the Chinese continued to import *wootz* steel from India and Persia, and one suspects that their own co-fusion steel seldom attained the quality desired.

The properties of steel can be greatly changed by heating and sudden quenching, followed by a gentle heating process known as annealing. Although these processes were common practice in mediaeval Europe, little is known about how they originated since, without written records, one can only be certain of their use by examining metallographic sections of early iron objects. As this involves a certain amount of damage to remove the section, little work has been done in this field.

Top left: drawing of a forge in France. The water wheel drives the forging hammers.

Left: the first all-iron bridge, built over the River Severn at Coalbrookdale, England, by Abraham Darby in 1779.

The lathe

According to legend, the lathe was invented by a Greek, Theodorus of Samos, in the seventh century BC, but the first direct historical evidence is an Egyptian tomb painting depicting a lathe in use in about 300 BC. In this picture the work-piece is shown held between the ends of a pair of horizontal spindles, while the turner holds his chisel against a support. The work-piece is made to rotate by a second craftsman who holds the free ends of a cord that has been wrapped around one spindle. By tugging the cord first with one hand and then with the other, the work-piece is made to rotate towards and away from the chisel edge. The turning process was thus intermittent, it being possible to cut only when the work-piece was revolving towards the chisel edge, and this was a feature of all early lathes. The return movement may have been useful for removing the wood-chip.

The development of the lathe has to be viewed in the context of its period, for from the seventh century onwards in the Eastern Mediterranean countries more and more machines were being built. These were largely made of wood, and spindles, pulley wheels and cog-wheels were essential parts of their construction, all calling for manufacture on a lathe. The principle of the lathe was probably suggested by the potter's wheel.

Although the lathe was primarily a wood-turning tool, it was adapted to many other uses. The softer stones such as calcite and alabaster were turned into bowls and vases, for example. Furthermore, by fitting a circular metal disc on the end of the driving shaft the stoneworker had a lapidary's wheel. Fed with fine sand and water, the edge of the lap wheel could be used for cutting either stones or glass, while the flat face of the wheel could be used for polishing. The tombstone of one Roman lapidary shows his wheel rotated, not

MARGARET MURRAY

MICHAEL HOLFORD LIBRARY

Top right: the bow drill is an ancient device, but is still to be seen in some parts of the world. Here an Indian makes beads from ivory and wood in Khartoum.

Below right: this is part of a treadle lathe of the late 18th century. A rope would have linked the two wheels.

by another man, but by a bow, the string of which passes around the drive spindle. The device could thus be used by a single operator.

In these early periods the lathe was not used for turning metal, because it was not possible to make sufficiently hard cutting tools. The lathe, however, was further adapted as a means of shaping metal bowls. A sheet of metal was attached to the end of the spindle, and as a result of pressing heavily on it with a burnishing tool, such as a polished agate stone, the metal was gradually forced into hemispherical shape. Bronze bowls made in this way are a fairly common feature of the late Roman Empire.

The lathe underwent no further improvements until the twelfth century when a new method of drive was introduced. A springy sapling was attached horizontally to a vertical post by one end so that the free end was positioned immediately above the lathe. From the free end a cord was carried down, around the spindle, and attached to a pedal. By pressing down on the pedal the turner made the work-piece rotate towards his chisel; by releasing his foot-pressure the sapling, acting as a spring, returned the work-piece to the starting position. The pole-lathe, as it was called, allowed the craftsman to use both hands to hold the work tool and, since greater force can be exerted by the leg than by the arm, larger pieces could be turned. This primitive form of lathe was still in use amongst the chair bodgers (a name used for craftsmen who made turned chair legs) in England at the beginning of this century.

A change in design that allowed the lathe to be used continuously, rotation of the spindle always being in the same direction, appears to have taken place in the early fifteenth century, when a simple form of belt drive was introduced. The energy was supplied by a large wheel, known as the great wheel, made to rotate by a pair of hand operated crank handles. From the circumference of this wheel a cross-over belt drive was carried round the spindle of the lathe. This form of drive was clearly inspired by the recently introduced spinning wheel, a device developed in the Far East, but it did not prove to be popular since, despite its high output, it demanded more manpower than earlier forms of the lathe. It is not until the Renaissance that one comes across a lathe with a flywheel and crank-and-pedal drive, as sketched by Leonardo da Vinci, that was both efficient and could be used by a single operator.

MICHAEL HOLFORD LIBRARY

Left: part of a pole lathe, made in about 1800. The rope is attached to a treadle at the bottom and to a springy sapling at the top; in this case it passes around the workpiece itself.

Below: a 17th century German woodcut of a pole lathe in use.

CR J F STENT

Plumbing

In the New Stone Age, between 9000 and 4000 BC, when settlements were small, springs and wells could provide all the water required by a community. After this period, however, large towns began to be developed in the Near East and water had to be brought from reservoirs and springs, often at considerable distances from the settlements. Usually water was carried in channels on conduits made of stone and fitted with covers to prevent evaporation and pollution. Many such conduits are known from cities of the Near East dating from 4000 BC onwards. This means of transporting water reached its height of development under the Roman Empire, and many Roman aqueducts are still largely intact today.

The aqueduct, however, had many disadvantages. Although water could be brought to storage tanks in the town, it was a cumbersome way to distribute water to individual buildings. Another disadvantage of an aqueduct was that water had to flow down a constant, gentle gradient between the source and the consumer: a system of siphoning water down and up hills could not be used in conjunction with an aqueduct. Finally, sewage water had to be disposed of, and since the covers of conduits were apt to become dislodged, they did not provide the best possible sewage system.

The earliest known water pipes are in the ancient cities of the Near East and the Punjab. The pipes were made of relatively short lengths of earthenware, often flanged at the ends to make it easier to join sections together with bitumen. Even at this early date the pipes appear to have been made to standard sizes, and provided water not only for baths but also for fountains and water closets. Similar earthenware pipes were in use more than a thousand years later at Knossos in Crete, but here the lengths were tapered so that the narrow end of one could fit into the wider end of the next, while a pair of loop handles was attached to each end of the pipe allowing the lengths to be tied together with cords.

Below: the system set up by the Romans at Hama, Syria, for water supply, is still in use, though now extensively renovated.

Bottom: earthenware water pipe, still to be seen in the ruins of Troy, which was inhabited from about 3000 BC to 300 AD.

INTERNATIONALES BILDARCHIV

NORMA SCHWITTER

Above: part of a Carthaginian drainage system. Based on Carthage, now in Tunisia, this empire spread over the North African coastline and rivalled the Romans.

The earliest known metal water pipes are from Egypt and date back to about 2000 BC. Each length of piping was made from a rectangular sheet of copper folded over and joined to form a cylinder with a bore of about $\frac{3}{8}$ inch (9 mm), 18 inches (45 cm) long. Copper, however, was expensive and it was not until lead became readily available after 1000 BC, as a by-product of extracting silver from galena ore, that metal pipes began to be used in quantity. They were made from sheet lead rolled into cylinders and jointed with tin solder, and in ancient Rome such piping became completely standardized as five inch, eight inch and ten inch (13, 20 and 25 cm) gauges according to the width of sheet rolled to form the pipe.

Piping was used mostly to distribute water within cities, but in some cases it was used to pipe water from the source. In the Phoenician city of Sidon, for instance, water was brought a distance of 15 miles (24 km) in earthenware pipes, while bronze pipes were used to bring water from the mainland to the island of Tyre. In the Greek city of Pergamom in Asia Minor water was brought by a siphon from a source on an adjacent hill some 90 feet above the city, the pipeline sinking into and rising again from an intervening valley 500 feet (150 m) below the town.

Perhaps nowhere in the ancient world was there more extravagant use of lead piping than in Rome. Although the water was brought into the city by aqueducts, it was distributed to households and public fountains by lead pipes connected to distribution tanks, private users paying taxes according to the size of the bore of the pipe. Even by today's standards, huge quantities of water were consumed. In 100 AD water was being used at the daily rate of three hundred gallons per head of population —about six times as much per head than is consumed in most modern cities.

Undoubtedly a great deal of water ran to waste in ancient cities, for even in well appointed Roman houses stopcocks and taps [faucets] were a rarity. At Pompeii, for example, a plumbed water heating system fitted with taps was unearthed, although such installations must have been a luxury that only the wealthy could afford.

Elsewhere in the Roman Empire, especially in France, Germany and Britain, water was often transported in wooden pipes made by hollowing lengths of log. The pipes were joined by means of flanged iron collars, and it is often these alone which have survived to show the presence of a Roman pipeline. Even in these outposts of the Empire, however, the larger Roman villas had plumbed water supplies to baths and fountains. So advanced were these plumbing systems that it was not until the 19th century that any improvements appeared.

RONAN PICTURE LIBRARY

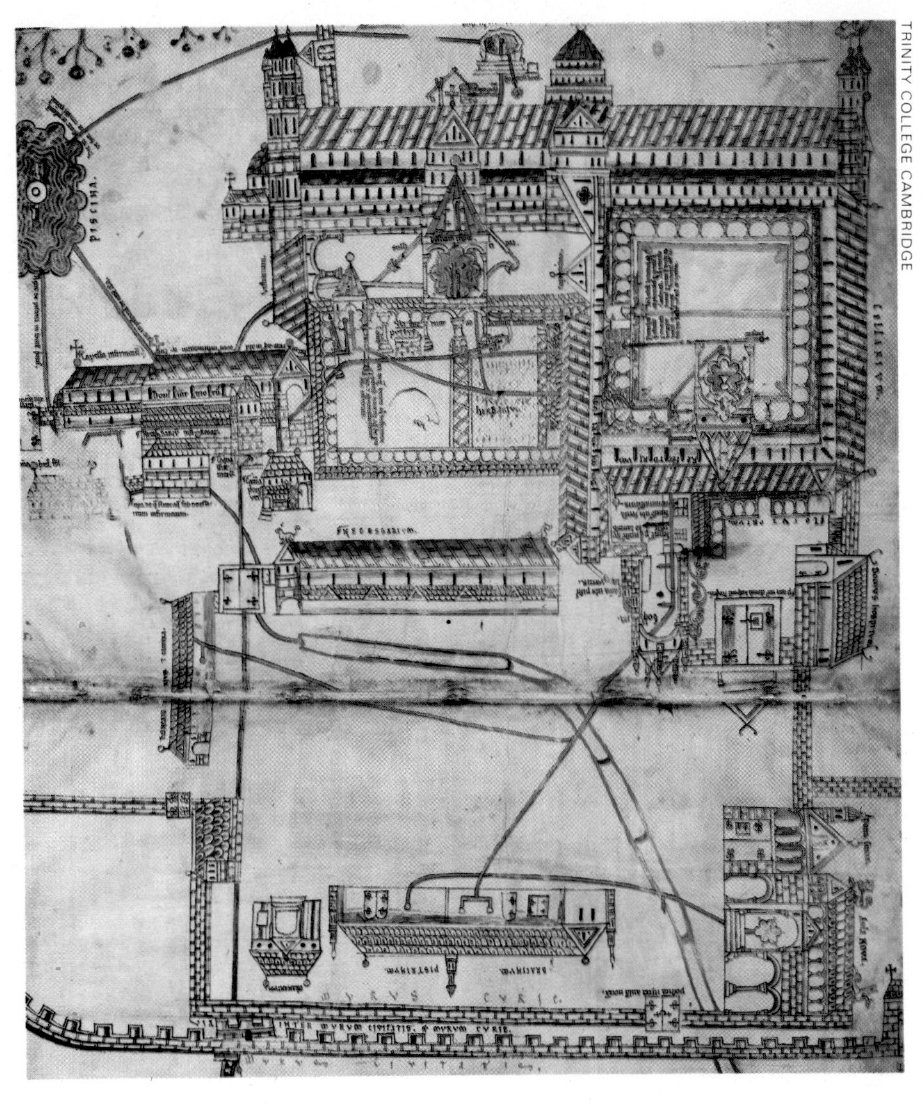

TRINITY COLLEGE CAMBRIDGE

Top: a drill for boring wooden pipes, shown in a book of 1664. The rotary power was provided by a waterwheel; a fairly narrow hole was bored using the augers marked E. Pipes like this became common in mediaeval Europe.

Right: 12th century plan of Canterbury Cathedral: water is green and pipes red.

Pottery

In the Ice Age, hunters made animal models which they baked hard in the fire. This discovery that clay could be fired to make a durable material was probably made several times, since pottery is known from both the Old and New Worlds. As far as Europe is concerned, the techniques of pottery making spread from the region of their discovery in the Near East, during the New Stone Age from about 7000 BC onwards, where mixtures of clay and sand had long been used as materials for building.

The earliest pots known were shaped either by pressing a slab of sandy clay over a rounded shape such as a gourd, or by adding successive rings of clay to build up the sides of the pot. They were usually air dried before firing to prevent cracking. When dry they were fired either in the domestic hearth or in a small bonfire. Soot and ash have left dark blotches on the surface of such pots, which are usually very porous.

One technique for reducing the porosity was *burnishing*. Using a pebble, the surface of the pot was rubbed to a smooth finish while the clay was still damp. This practice closed the pores and produced a characteristic shine on the fired ware.

Early in the New Stone Age a method of decorating pottery was developed which involved the use of differently coloured clays or *slips* painted on to the surface of the damp vessels. If the decoration was not to be spoilt during firing it now became necessary to separate the pots from the fire, and from various archaeological sites the gradual development of the KILN can be traced. At first the fire was lit below a clay grid upon which the pottery was set; a low wall was built up around the grid to contain the vessels and retain the heat; the wall was made progressively taller so that more pots could be fired at a time; and finally this upper structure was built as a dome with a small vent at the top.

At the same time that the kiln was being developed methods of shaping were changing too. A simple turntable enabled vessels to be modelled more easily. It was soon discovered that by making the turntable spin rapidly it was possible to make thinner and more

BRITISH MUSEUM/MICHAEL HOLFORD

STAATSBIBLIOTHEK BERLIN

ARCHAEOLOGICAL MUSEUM—PHOTO: C. M. DIXON

ROYAL SCOTTISH MUSEUM—PHOTO: C. M. DIXON

Above left: Neolithic pottery found at Camp de Chassey in France. These pots were reconstructed from fragments, as whole pots are seldom found intact.

Above right: a clay figurine of a water buffalo, found in the Hacilar region of Turkey. It is believed to have been made some time around 6000 BC

Far left: a 'Kamares' style vase from Phaistos, on the island of Crete. It was made between 2000 and 1700 BC by the Minoans, who ruled Crete for 1500 years.

Left: an Aztec baby's rattle made in Mexico in the 15th century AD, during the last phase of Aztec civilization before the Spanish invasion of 1519.

evenly shaped vessels from a single lump of clay placed in the centre of the wheel.

The development of the kiln and the potter's wheel brought other changes. It was now possible to use a finer clay for making pottery, and this had to be specially prepared by mixing the clay with water and allowing it to settle in large troughs. The coarser clay sinks to the bottom of the trough leaving the finer particles above it. After much of the water has evaporated the fine putty-like clay on top is collected. Thus, by the end of the New Stone Age, about 3000 BC, the whole process had been transformed. It was now no longer a craft carried out by individual households, but a specialized industry.

Few changes were made in the subsequent two thousand years, and it was not until about 500 BC that it was realized, probably by an Athenian potter, that by making the wheel taller it could be rotated by kicking the lower part with one's foot. This gave the potter a greater control over the speed of the wheel and more rapid production. It was not, however, until the Roman period that a means of mass-production was evolved in which the use of the wheel became subordinate to the use of moulds. Samian pottery, made in various centres in Italy and Gaul (France), was formed by pressing the clay into moulds, many of which carried ornate patterns in negative relief. These moulds were then set briefly on a potter's wheel and spun so that the inner surface of the vessel could be smoothed, a process similar to that used in industrial potteries today.

Before 200 AD very little glazed pottery was produced, largely because of the difficulty of making a glaze that did not peel away from the surface of pottery. One solution to this problem was found in Egypt and Mesopotamia a little before 1000 BC, and this was to use finely ground white sand and soda in place of clay as a material from which to manufacture vessels. When fired such pottery automatically had a glassy surface. This material was difficult to shape and expensive to prepare, and as a result were comparatively rare.

An alternative solution to the problem lay in covering the unfired pots with a coating of lead oxide. On firing the lead combined with the sand in the clay to form a glaze which bonded firmly to the surface of the pot. Curiously this method appears to have been adopted by the Chinese and by many potters in the Roman Empire at much the same time, about 100 AD. Thereafter glazes containing soda or lead, or mixtures of the two, became common in Europe and Asia, and are still widely used today.

SONIA HALLIDAY

Left: a Chinese engraving of 1811 AD, showing workers making porcelain ware. The man on the right is shaping the vessels from clay, on a wheel which the man holding on to the rope is turning with his foot. The finished pot is dried in the shade, and then fired in the oven. It is taken hot from the oven and plunged into water containing quartzose sand, which forms the glaze. Then it is fired again, decorated, and finally dried gently in the oven.

Below: making pots by hand, without using a wheel, in a Nigerian village.

RONAN PICTURE LIBRARY

The plough

Ever since antiquity the plough has been regarded as one of mankind's most beneficial inventions. In many societies its origins in folklore were attributed to a god or divine king. Before the invention of the plough cultivating the land was extremely arduous work: the soil had to be tediously broken up with hoes and mallets before the grain could be planted.

There is evidence that the first ploughs came into use in Egypt and Mesopotamia in about 4000 BC. In both countries a picture of a plough was used as a word-sign in early forms of writing. The construction was very simple: a forked branch was lashed to a pole. The junction of the two forks acted as the ploughshare which dug into the soil. The ends of the forks became the handles, while the plough was pulled along by the pole.

Probably this kind of plough was first dragged by humans, although the Mesopotamian word-sign for a plough already depicted a yoke for a pair of oxen. From then onwards a pair of oxen became the standard team for pulling the plough.

This simple, ox-drawn plough remained in use in Egypt until recent times. In Mesopotamia, however, a number of changes were made before 2000 BC which enabled ploughing and sowing to take place simultaneously. The forked stick was replaced by a pair of handles set into a pointed wooden block which made a bigger share. A vertical hole was then drilled through the block into which a tube was fixed. At the top of the tube was a funnel through which the seed was then poured for planting. In this way, sowing could be done at the same time as ploughing. Such ploughs are still in use in remote parts of the Middle East today.

STAATSBIBLIOTHEK BERLIN

EUROPEAN COLOR LIBRARY

SNARK INTERNATIONAL

Top: it was in the Neolithic period that men first seriously turned to agriculture. This French rock illustration shows that the basic design of animal drawn plough was a very early invention.

Centre: an Egyptian plough shown in the tomb of Sennedjen, about 1250 BC.

Bottom: this plough, shown in the Luttrell Psalter of about 1338, shows an English plough with a large mould-board.

In prehistoric Europe a substantially different type of plough was in use by 2000 BC. Again the basic design depended upon a forked branch, but in this case one fork was cut short and sharpened to form the share, while the other was left long to serve as the pole. A single handle, fixed to the share and pole, had the advantage of leaving one hand free for the ploughman to goad the ox team. At the time of the Roman conquest in the first century BC, various forms of this kind of plough were in use throughout Europe.

In the Mediterranean countries from 1000 BC onwards there developed what appears to be a hybrid between these first two types of ploughs. In early Greek illustrations, for example, the plough had a heavy, pointed wooden share, like those of Mesopotamia, and into it were set the pole and a single handle similar to the early European plough. After about 800 BC the share was sheathed with an iron point.

It was this kind of hybrid plough which the Romans introduced into Europe, but it had many disadvantages especially when used on the heavy, damp soils of Europe. The share tended to either clog or skid over the surface of the soil which prevented it from making a furrow. To overcome this difficulty the Romans fixed a sharp blade, the *coulter,* to the pole ahead of the share. In this way, the coulter opened up the ground for the share.

In spite of the coulter, all these early types of ploughs did no more than make a groove in the soil. None of them could turn the soil over efficiently. This required a mould-board, a twisted plank set behind the share, which would slip under the cut soil and force it over as the plough passed.

The invention of the mould-board is generally believed to have been the work of the Slavs who lived in the region of the Pripet Marshes, today the western frontier of Soviet Russia. It was from this region that other peoples, both in the West and Far East adopted the mould-board. A pair of wheels was often added to support the weight of the plough, while the ox team was usually increased to four animals. The mould-board type of plough was used throughout Europe during the Middle Ages.

The effect of this plough on European agriculture was enormous. Heavy, damp soils, which up to then were unworkable, could now be opened up. By ploughing up and down hill, leaving channels between strips of land, excess water could be drained off. This system of ploughing has left its mark in several ways. Fields covered with parallel ridges and furrows, still to be seen from the air, remains of such ploughing; while the acre, a plot of land measuring 22 yards (20 m) by 220 yards (200 m), the standard measure of land, was in the Middle Ages regarded as the area which a team could plough in a single day.

Below: wheeled ploughs were known in Roman times, though they were rare in Europe before the 11th century, which is the approximate date of this illustration. The wheel helped to support a heavy plough, though they were by no means essential and never became widespread.

Bottom: in many parts of the world, the traditional ploughing methods are still used. This is a plough with a mould-board from present-day Tunisia.

SCALA

SONIA HALLIDAY

Salt mining

Salt was needed by early man for many different purposes. Quite apart from cooking, salt was used extensively for curing skins and for preserving foodstuffs. The ancient Greek historian, Herodotus, for example, described the salting of fish by the people living around the Dnieper estuary, while the Greek writer, Strabo, praised the quality of pork salted in Gaul. To a lesser extent salt was required for a number of industrial processes such as dyeing and soap making.

In the relatively hot countries of the Mediterranean and the Middle East the preparation of crystalline salt from sea water or the brine from salt lakes presented few problems since it was easy to obtain by evaporation. A common method was to cut tanks or pans in rocks close to the sea shore. The pans were filled with water; as it evaporated more was added until, finally, the concentrated brine was allowed to dry out completely and the salt crystals were removed. This method is still in use in many places such as Malta where the pans are certainly mediaeval in origin, if not earlier. In other areas, where there was no suitable rock for cutting pans, they were made of clay but used in a similar manner.

In temperate Europe natural evaporation was normally too slow to make this process practicable, and the brine had to be heated. One system was to put the brine in conically shaped earthenware pots supported on clay stands. A slow burning fire was kindled beneath them, and as a result of continuously topping up the vessels with brine they eventually became filled with salt, which could remain in the pot for transportation. Alternatively a series of bricks might be heated from below, and the brine allowed to drip slowly on to their upper surfaces. As a result, crystals of salt were formed on their surfaces and could be scraped off.

Various views of a European salt mine, showing how the salt is cut out in blocks, forming 'rooms' with large pillars between them to support the ceiling.

In central Europe, the transportation of salt from the coasts would have been an expensive undertaking. It is hardly surprising, therefore, that both salt springs and rock salt were exploited even in prehistoric times. Remains of such early salt works have been found at Sielle in Lorraine and at Halle on the Saale in central Germany, in both cases the source being natural salt springs.

The deposits of solid rock salt of the eastern Alps were certainly being mined by 500 BC; the remains of such workings have been found at Hallein, Hallstatt, Salzburg and Dürrnberg. Because the miners' abandoned tools of wood and leather have been perfectly preserved by the salt, a great deal is known about the way in which they worked. Previously, mining for other minerals such as the ores of metals had been done by opencast methods or by digging shallow pits. In neither case could the mine be kept sufficiently dry to prevent the salt from becoming dissolved. Indeed, in some cases the miners deliberately allowed the salt to become dissolved and later

MARY EVANS PICTURE LIBRARY

recovered it by evaporation. On the other hand, to prevent the mine from becoming damp the miners learned to drive a passage, or adit, into the hillside in order to get at the deposits of rock salt, and in so doing appear to have introduced adit mining into Europe.

In prehistoric China, as in Europe, sea water provided a source of salt near the coast, while in the mountainous western parts of the country it seems that rock salt was also being mined. More interesting, however, because of the techniques evolved by the Chinese for its extraction, was the exploitation of deep deposits of brine in Szechuan Province. By 200 AD a method had been worked out for sinking narrow bore wells, often to depths of 1800 ft (550 m). A tower of scaffolding was erected, with a pulley block suspended from its top. Over this ran a rope to one end of which was attached a length of heavy bamboo fitted with a substantial point made of a hard stone. By raising and dropping this primitive drill it slowly pulverized the rock, the material being removed by a hollow bamboo scoop passed down the bore hole from time to time. Slow as this method of drilling must appear to us today—one Chinese historian records a rate of 800 strokes per hour—it must be remembered that the prospectors for oil in the last century used almost identical equipment.

Once the deposit of brine had been reached, the bore was lined with lengths of bamboo wrapped with canvas and mastic—a natural resin—to prevent the well from becoming choked, and the brine was winched to the surface in hollow bamboo containers. The brine was then transferred to shallow heated pans and it was thus evaporated to give crystalline salt.

By the twelfth century the Chinese had considerably improved upon this technique of recovering brine from deep deposits, and one bore of a depth of 3500 ft (1068 m) is recorded. The working of the drill and lifting gear was now done by an ox-drawn capstan. More remarkable, however, was the fact that in drilling for brine the salt workers had not infrequently come across pockets of natural gas. The gas was tapped by means of tubes of bamboo to the brine heating plant where it was used to warm the salt pans. This appears to be the first recorded industrial use of natural gas, although difficulties of transport and storage prevented it being used for other enterprises.

RONAN PICTURE LIBRARY

Above: salt mines at Cracow, Poland, in 1820. The miners are starting to work their way into the hillside; the excavation will later develop into a long passage to reach the interior of the salt deposits.

Below: aerial view of salt pans in Aden, where seawater is evaporated to leave salt, which is then scraped out of the pan. This method requires a very hot climate to be economically feasible.

SPECTRUM

Canals and locks

Canals, designed only for the purpose of irrigation and water supply, were a common feature of all the ancient civilizations of the West—Egypt, Mesopotamia, Greece and Rome—but the idea of creating artificial waterways remained largely embryonic. There was indeed little need for canals. In Europe the seas and the large rivers provided adequately for the limited flow of goods, while in the Middle East the general lack of water made such enterprises unworkable. A notable exception, however, was the canal designed to join the east bank of the Nile Delta with the Red Sea at Suez. This canal was begun in the reign of the Pharaoh Necho in about 600 BC, and according to the ancient historian, Herodotus, was wide enough to allow two ships to pass. In the late Roman Empire plans were drawn up to construct a canal across the Suez Isthmus but the technical difficulties proved too great and it was abandoned. Even as late as 800 AD the Frankish Emperor, Charlemagne, laid plans to join the Rivers Main and Danube, but only two miles of this canal were dug before the project was given up.

In early China the situation was quite different. The three great rivers—the Yellow, Yangtse and West Rivers—flow largely from west to east, while the economic situation needed a system of transport that ran from the food-producing south to the north. Furthermore the rivers, especially the Yellow River, were prone to violent flooding. Early Chinese canal construction was designed, therefore, both to provide a system of transport and to relieve floodwater.

JOHN FREEMAN

Right: an early 17th century European lock. The lower lock gates are shown opening in the wrong direction.

Below: this engraving of the 1840s shows Chinese junks still using an inclined plane, the earliest form of lock.

MARY EVANS PICTURE LIBRARY/COLOUR: SALLY SLIGHT

The first steps in creating what was ultimately to become the Great Canal were taken in 350 AD with the opening of a waterway between the Yangtse River and the Huai River some 120 miles to the north. This canal, which ran roughly parallel to the coast, had to rise about fifty feet as it ran north. This was made possible by the introduction of locks. Originally the Chinese had overcome the problem of water transport in rivers where there were rapids by building a weir with a ramp alongside, boats having to be manhandled up and down the incline. This operation consumed a great deal of time and energy, and by 100 AD Chinese engineers had found an alternative solution. The ramp was now incorporated into the bed of the canal, and a gate was placed at the lower end of the incline. This gate took the form of two stone pillars, one set on each side of the canal. The posts were grooved down their inside lengths, and into the grooves were fitted the ends of a number of horizontal logs which served to hold back the water to the higher level. The logs could be raised by means of ropes attached to their ends, while at a later date the logs were joined to form a single gate that could be raised or lowered like the blade of a guillotine.

This simple form of lock, usually referred to as a *flash lock*, was normally provided with a capstan at the higher end so that boats moving to the upper level could be winched up the incline down which water was pouring at a considerable speed. Boats moving to the lower level had to 'shoot the rapids' down the ramp, often at great risk. Even so, despite improvements to be mentioned below, the flash lock has remained a common element in the smaller canals of modern China.

By 600 AD the Great Canal had been extended south from the Yangtse River to Hangchow, a distance of about 220 miles, and in the following three centuries some ninety canals were constructed throughout China, while in the period from 950 to 1280 AD no less than five hundred such projects were completed.

There seems little doubt that this spate of canal building was at least in part engendered by the introduction of a new and more convenient form of lock, the earliest record of which goes back to 983 AD. The Chinese engineers had realized that by using two gates set at some distance apart, rather than a single lock gate, the hazards and labour of the flash lock could be largely overcome. Boats could now enter the 'pound' created between the two gates, and the water level could be raised or lowered gently, so doing away with the risks of the flash lock. The *pound lock* now made possible the construction of the final and major part of the Great Canal. Between 1280 and 1300 the canal was extended north to Peking to provide a waterway just over a thousand miles long and sufficiently deep and large to accept small sea going ships.

When canal building began in mediaeval Europe—principally in Holland where the programme of land reclamation made the use of waterways a logical form of transport—the development of locks followed a similar pattern to that of China. Thus by the eleventh century the Dutch had evolved, quite independently of events in China, a flash lock with a guillotine gate. By the late fourteenth century the pound lock had been developed and was quite common. The gates were still of the guillotine type, but were provided with massive lead counterweights and sluiced (fitted with small gates) to allow the gradual filling and emptying of the pound. In Italy, too, the construction of canals became fairly common in the late Middle Ages, especially in the area of Milan where a suitable form of transport was required to handle such bulky goods as marble. It is not surprising, therefore, that the first record of hinged gates to locks, such as are normal on Britain's canals today, are to be found in a sketch made by Leonardo da Vinci made at the end of the fifteenth century.

DAILY TELEGRAPH COLOUR LIBRARY

PHOTRI

Above left: the 'staircase' of locks built in 1825 on the Grand Union Canal at Foxton, Leicestershire, is still in use.

Below left: the Erie Canal, famous in American folklore, was opened between New York and Lake Erie in 1825.

THE MIDDLE AGES & THE RENAISSANCE

In the Middle Ages men continued to be technically innovative. The Renaissance meant a re-birth of Man's confidence in his own worth. The map on these pages is a Dutch map of America, originally engraved in 1608 and corrected by Willem Janzoon Blaeu. This copy of 1655 is in the National Maritime Museum, Greenwich, England. The Age of Discovery signified a new way of looking at the world, used by a new kind of craftsman: the scientist.

AMERICÆ TABULA, auct: G. BB

The clock

The mechanical clock was evolved in the later Middle Ages to meet two rather different demands. In the first place, there was a need for a simple timepiece to record the hours for the regulation of work and church services. On the other hand, astronomers required a machine to drive their astrolabes by which they could study the relative positions of the Sun, Moon and stars. In the early Middle Ages the first of these functions was performed by some form of water clock, or clepsydra, in which a vessel either emptied itself or was filled with water over a set period of time. At the end of each period an attendant, the 'watch', had to recharge the water clock, and its accuracy depended upon his alertness. Even by 800 AD, however, the Emperor Charlemagne owned a water clock that automatically released a bronze ball into a bowl at every hour, a primitive form of striker. Over the next three centuries striking mechanisms were to become even more elaborate, frequently taking the form of a mechanical figure that struck a bell.

The striking train of early water clocks was normally made to work by a falling weight and cord drive, the cord being carried around a spindle that operated the gears, levers and pulleys that moved the figures. At the same time it was realized that a weight and cord drive could be made to move both astrolabes and timepieces if only some means could be found to regulate the speed at which the weight fell; since there was already a tradition of using the flow of water in clocks it was natural that a hydraulic regulator came to be employed. In a typical example the weight was prevented from falling by a ratchet wheel and pawl. The pawl could be released by the free end of a balanced lever on the other end of which was fixed a container. Water dripped into the container until it was sufficiently heavy to move the lever, so tripping the pawl and allowing the ratchet wheel to move forward a single space, while the container emptied itself and the lever arm returned to its original position. By 1200 AD this basic type of regulator was in use in various different forms in China, the countries of Islam and Europe, and clocks were in sufficient demand for a city such as Cologne to have its

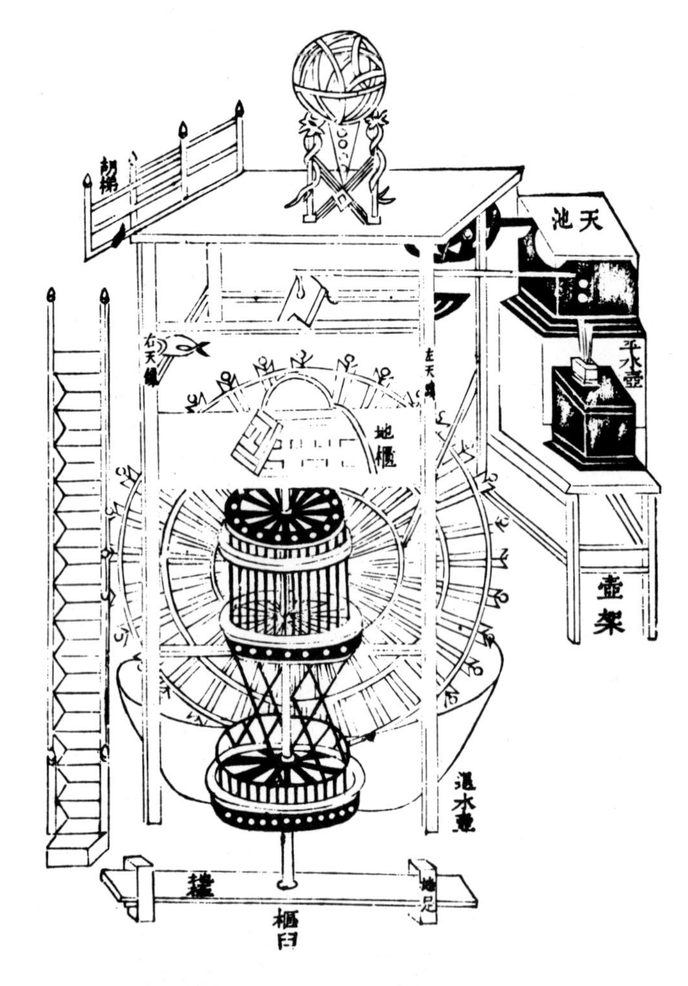

Above: a general view of Su Sung's clock tower, from Hsian Hsiang Fa Yao, c. 1094. It is impossible to establish with certainty, but Su Sung's was apparently an entirely mechanical escapement, which precedes similar development in European timekeeping by about two hundred years. One source sets the date of the invention even earlier, giving credit to Yi Hsing in AD 725. The first mechanical escapements in Europe date from the late 13th century. One of the oldest still in use is at Wells Cathedral; it dates from 1392.

Left: an illustration of an elephant carrying a mechanical clock, from 'The Book of Knowledge of Ingenious Mechanical Devices', compiled by al-Jazari in 13th century Baghdad for the Sultan of Amid.

own guild of clock makers.

Water operated escapements of this kind however, suffered from many disadvantages. In cold climates the water would freeze, and in hot countries it would evaporate. Furthermore, the aperture through which the water dripped eventually either became enlarged by erosion or clogged by deposits from hard water, so making the instrument increasingly less accurate. In 1271 AD a monk, Robert the Englishman, writing about clocks of his day, realized that a purely mechanical escapement would be preferable, and remarked that a number of people were at that time working on the problem, for which by then no solution had been found. A few years later some unknown engineer either in northern Italy or France hit upon the answer with the invention of the *verge-and-foliot* or *verge-and-palette* escapement.

The verge-and-foliot escapement took the form of a vertical rod (the verge) on which were set two small metal flags (the foliots or palettes), the verge being suspended on a short length of cord or sinew while its lower end was held in a bearing. The verge could thus twist first in one direction and then in the other, and as it did so the foliots alternately engaged with and released the teeth of the ratchet wheel. The speed at which the verge oscillated was controlled by a short bar fixed at right-angles to the top of it. On each side of this bar was suspended a small weight. By moving the weights towards the verge the speed of oscillation was increased; by shifting the weights outwards the rate was decreased. The speed of the clock was thus fairly easily adjusted.

In some early Italian clocks, however, the verge was fitted with a wheel in place of the cross-bar, small weights being suspended from its spokes, again for the purpose of adjustment. In 1364 AD an Italian professor of astronomy, Giovanni de' Dondi, perfected such a clock, the details of which he had been working on for sixteen years. Indeed, during the fourteenth century it became a matter of civic pride for most communities to become the owners of a clock, usually housed in a conspicuous place such as the tower of the local church.

By 1400 AD coiled springs were beginning to be used in door locks, and shortly afterwards in the firing mechanism of some flintlock handguns, which were actuated by a spring-operated hammer. In 1430 AD coiled springs were first used to replace the weights of clocks. These allowed a far more compact form of time-piece, and one more suitable for travellers and navigators since they were less easily put out of adjustment by the movements of carriages and ships. The verge-and-foliot escapement, however, remained in use until about 1650 AD, when the swinging pendulum replaced the oscillating bar in many clocks, so giving the type of standing clock still to be seen today.

Below: a French illumination from 1454. A clock stands on Earth but with its works, dial and bell in Heaven. In the four corners of the dial, the four symbols of the gospels: angel, eagle, lion, bull. The goddess Attemprance, dressed in red, winds the clock.

BODLEIAN LIBRARY

Below: a table clock signed Nicholas Lemaindre, Blois, 1619. The clock is $7\frac{1}{4}$ inches high and has a mainspring movement with a fusee. The fusee is a tapered barrel with gut or a chain wound on it which keeps mainspring tension constant as it unwinds.

COOPER-BRIDGEMAN LIBRARY

Crank handles

A crank handle is a handle on the end of an arm at right angles to a shaft which is to be turned. The alternative, a spoked wheel, can be awkward to use, but one's whole body weight can be put into turning a crank, and it converts back and forth (reciprocating) motion into rotation or vice versa. The total absence of any form of crank handle is a surprising feature of all early machinery. A careful study of Roman and early Chinese machines shows that whenever a shaft had to be turned it was fitted either with spoke handles or a wheel.

The earliest known instance of a crank handle is to be seen in a manuscript drawing of the early ninth century AD, made near Reims, in which the crank handle is being used to turn a grinding wheel for sharpening swords. Even four centuries later the crank handle was still uncommon and its uses limited. In one of the thirteenth century glass windows of York Minster, craftsmen are depicted turning the mould core for a bell on a spindle fitted with a crank handle, but at the same date hoists and winches were still equipped with spoke handles.

In Europe it was not until the first half of the fourteenth century that the crank handle began to become more widely used, when the double crank, one which bends back again to the line of the shaft as typified by the carpenter's brace, first appeared. From then onwards the double crank was to become quite common, both as a handle and as a crankshaft. Thus a small mill might be driven by a crankshaft to which was fixed a handle alternately pushed and pulled by the operator. Probably, however, the crankshaft was not a European invention since an Arabic handbook of engineering depicts such a device in the early thirteenth century—a hundred years before it appeared in Europe.

Why the craftsmen and engineers of the Middle Ages were so slow to adopt the crank handle demands some explanation. First, very few mediaeval machines

Above right: crank handle used to work a multiple crane; a print of 1578.

Below right: crank on a hugely complicated geared winch for hauling building stone along a tramway, 1588.

incorporated a rapidly moving element, and those that did were normally driven by wind or water power through gearing. Furthermore, by the end of the twelfth century engineers had learned how to convert the rotary movement of the water wheel to linear movement by attaching cams to the drive shaft. Hence a pivoted hammer (a trip hammer) might be raised by a cam engaging with the end of its beam, then allowed to fall under its own weight.

Another obstacle to the introduction of the crank handle was that its use normally required two men, one to turn the handle and the other to use the machine (such as a grindstone or lathe). In the twelfth century this problem had been overcome by the employment of the pedal-and-spring drive system. A cord was attached to the free end of a springy sapling fixed above the equipment. The cord was carried around the drive spindle and down to a pedal. By pressing down on the pedal the spindle was made to rotate, and by releasing the pedal the spring made the spindle turn in the opposite direction. Although the movement of the spindle was a reciprocating one, at least the craftsman had both hands free to do his work. The pedal-and-spring became widely used in the Middle Ages for saws, hammers and looms as well as lathes.

Finally, in many pieces of equipment neither the crank handle nor the crankshaft drive could be used efficiently until it was realized that a flywheel was essential to provide sufficient momentum to carry the shaft over the two 'dead spots' of each rotation. The grindstone and the mould for a bell, both being heavy, acted as their own flywheels. This was not true, however, of a hand mill since the friction set up by the grain between the two stones was quite enough to offset any momentum.

By the fifteenth century, two quite distinct types of flywheel were being used in Europe. The first was no more than a large, heavy wheel attached to the shaft, as incorporated in Leonardo da Vinci's design for a lathe. The idea of using such a device may well have come from observing the effect of the large drive-wheel on contemporary spinning-wheels, or the kick-wheel used by potters, for example, where once they had been set in motion, such was their momentum that little further effort was needed to keep them going.

The second type of flywheel to appear at this time was a ball-and-chain system, usually attached only to vertical shafts. Four short chains were each linked at one end to the shaft, while to each free end was fixed a heavy ball. As the shaft was turned the balls flew outwards to provide the momentum. This device appears to have been an unparalleled invention of the period, but who the inventor was is quite unknown.

While the crankshaft and flywheel were to become increasingly used during the Renaissance, and were to pave the way for much of the machinery of the Industrial Revolution, their use remained restricted in the earlier periods to operations in which a steady speed was required. Thus, the Renaissance Italian potter Piccolopasso used a crank-operated mill to grind the materials of which his wares were made. At the same time the wheel on which he threw his pots, the speed of which he needed to vary, had a straight driveshaft with the potter kicking directly upon the wheel.

JOHN R FREEMAN & CO

A missed opportunity: this 1819 Rowlandson cartoon of the Prince Regent shows a crank used to drive a bicycle, but the idea was not actually put into practice for another 42 years.

Spectacles

Until the early fourteenth century those who were born with defective eyesight and the aged had no hope of being able to read or carry out any work that demanded good vision, for until then spectacles were unknown. Although fundamentally very simple, no eyeglasses could be made before the discovery of a number of techniques. Apart from a basic understanding of optics and the behaviour of lenses, it was essential to be able to produce clear glass and to know how to grind the lenses to the correct curvature.

As early as the tenth century the Arabs had made a special study of optics and diseases of the eye, always prevalent amongst desert dwelling people. Of these Islamic scholars the most outstanding was Ibn al-Haytham who worked in Egypt. In about 1040 he produced a treatise on optics in which, for the first time, the true function of the cornea (or lens) of the eye was described. Al-Haytham not only studied the anatomy of the eye but also the path of light as it was affected by mirrors and glass lenses, his workshop being equipped with lathes and lap-wheels for the production of his experimental materials. Strangely, however, al-Haytham never hit upon the idea of using lenses to correct faulty eyesight.

By the late twelfth century copies of al-Haytham's treatise on optics were widely distributed throughout the libraries of Islam, most importantly in Spain, then under Arab rule. It was here that his works were translated into Latin, then the common language of European scholars. Known to Europeans as Alhazan, al-Haytham's treatise formed the basis upon which many scholars worked. In England, for example, Grosseteste at Oxford University and his most brilliant scholar, Bacon, improved upon al-Haytham's theories.

At the same time many outstanding Arab scholars were attracted to the court of the Emperor Frederick II in Sicily. Here, too, the works of al-Haytham were translated into Latin. Thus, by the middle of the thirteenth century many European scholars had become familiar with the basic theories of optics, and a small number were producing their own experimental optical equipment. The manufacture, however, of a limited number of lenses for this kind of work was a very different matter from the production of large numbers of lenses as would be required for the widespread use of spectacles.

In the early part of the Middle Ages, the manufacture of high quality clear glass had been the virtual monopoly of the glass workers of Constantinople. The Byzantines had realized the importance of using only chemically pure ingredients to avoid the glass becoming

Below left: the first known picture of somebody wearing spectacles comes from a church fresco of 1352 by Tommaso da Modena, a monk. Early eyeglasses, unlike today, were not made to prescription.

Below right: 'a vain book collector wearing spectacles who dusts but does not read his books'—from a 1497 publication.

MARY EVANS PICTURE LIBRARY

NATIONAL LIBRARY OF MEDICINE

RADIO TIMES HULTON PICTURE LIBRARY

THE SCIENCE MUSEUM/PHOTO: DAVID KELLY

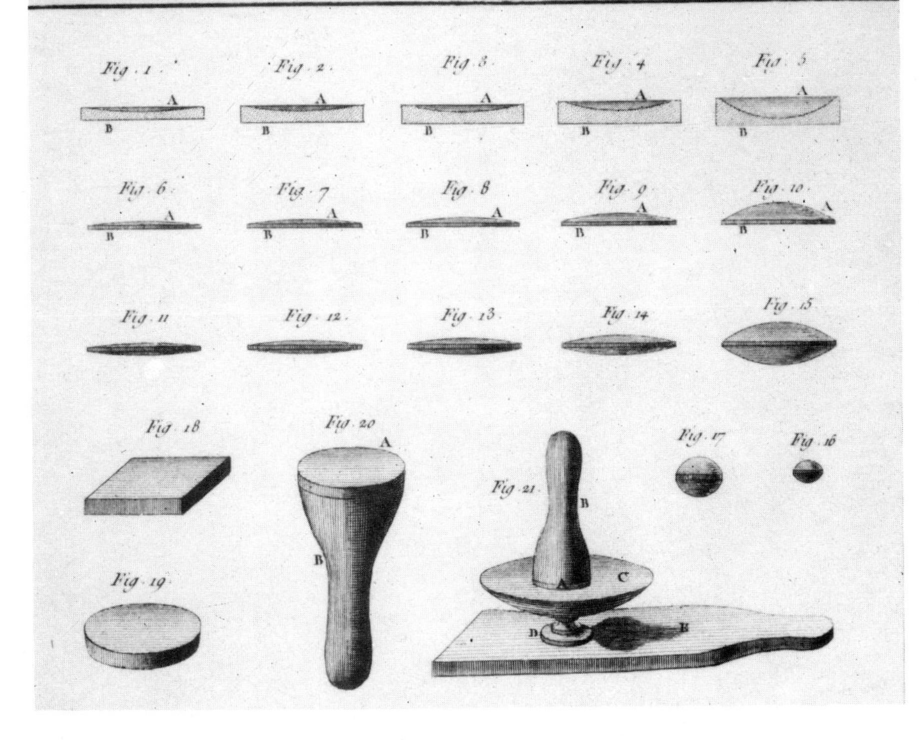

coloured, and the virtue of adding some lead to the glass composition to give the final product a degree of brilliance. They had, furthermore, become adept in the use of the lap-wheel for polishing and cutting their glasswares. This monopoly was finally broken when, during the Fourth Crusade (1204 AD), the Venetians directed their attack against Constantinople instead of the forces of Islam in the Holy Land. As a result of this outrage the Venetians were able to carry off a number of key workers in the glass industry, which enabled them to improve upon the quality and techniques of the established glass-works on the Islands of Venice and the adjoining mainland of northern Italy. By the middle of the thirteenth century Venetian glass had already gained an enviable reputation throughout most of Europe.

It was, thus, no accident that the first historical reference to spectacles occurs in Italy shortly after 1300 AD. It was here that the theoretical understanding of optics and the practical ability to produce suitable and sufficiently inexpensive lenses met, enabling the manufacture of spectacles to become a viable concern.

During the following century the use of eyeglasses spread slowly throughout the rest of Europe. Possibly because they still lacked the technical knowledge to make suitable glass, the use of spectacles appears to have spread less rapidly in the Islamic countries of the Near East. Thus it was that when the Portuguese and Dutch merchantmen usurped the Arab trade to the Far East by rounding the Cape of Good Hope during the sixteenth century, the Chinese had not yet learned of spectacles from Islamic traders. Like the Arabs the Chinese had certainly known of the use of lenses since the eighth century, but they, too, appear to have lacked the technical skill to produce adequately clear glass for their manufacture. Until the introduction of eyeglasses by the Europeans, the Chinese had been obliged to make do with magnifying glasses ground from pieces of clear quartz.

Top: a spectaclemaker's shop. Until the 15th century only spectacles with convex lenses, suited to the aged suffering from loss of accommodation, were available.

Left: an optical workshop and details of lenses from Diderot's 18th century encyclopaedia. Both concave and convex lenses were made. In the bottom corner a lens is ground using an abrasive.

The magnetic compass

There is little doubt that the magnetic compass was invented in China, but exactly when it was evolved, and when it was first used for navigational purposes, is uncertain. This is probably because the Chinese first used the magnetic compass as a method of foretelling future events. The first clear reference to the manufacture of a magnetic compass is in a Chinese encyclopaedia of 1040 AD in which the making of magnetic needles is described. The earliest record of its use at sea by Chinese mariners is in a report of 1115 AD.

Originally the Chinese compass took the form of a piece of lodestone (magnetic iron oxide) which was hollowed like the bowl of a spoon, and floated on water. The lodestone float, acted upon by the Earth's magnetic field, would always take up the same orientation along the north–south magnetic axis. This fact must have been known to Chinese augurs (fortune tellers), which, it is suspected, allowed them to 'fix' their predictions with some ease. In time the Chinese learned that iron could be magnetized, either by stroking it with lodestone or by heating it and allowing it to remain stationary while it cooled.

The original lodestone float was a cumbersome form of compass and the Chinese soon learned to use a wooden float, usually shaped like a fish, which incorporated either a piece of lodestone or a magnetized bar of iron. The first form of mariner's compass of which we know had the pointer, still shaped like a fish, balanced on a vertical pivot. This was, however, far from ideal as a navigational instrument in anything but very calm seas.

It is questionable how the magnetic compass came to be used in Europe and Islamic countries. The first known de-

MICHAEL HOLFORD LIBRARY / SCIENCE MUSEUM

MICHAEL HOLFORD LIBRARY / NATIONAL MARITIME MUSEUM

Above right: this Chinese compass, dating from the 19th century, has a central needle whose red end points south rather than north. It was not used for navigation but by geomancers, who directed the orientation of buildings or the burial of the dead. The card bears trigrams which symbolize directions.

Right: an early eighteenth century mariner's compass with a freely pivoted needle. Early compasses had no lettering but ornate pointers for north and east.

scription of a compass in Europe is in the writing of the Briton, Alexander Neckam, Abbot of Cirencester (1157–1217). Towards the end of the thirteenth century the mariner's compass was being widely used in Europe, Scandinavia and Iceland. There is no clear evidence that Islamic seamen were using the magnetic compass before 1232 AD, and the name given to it by the Arabs, *al Konbas*, is not an Arabic word, but probably borrowed from Italy. The earliest form of compass used by Islamic navigators was fish-shaped, like that of the Chinese. These facts have led most historians to suppose that Europeans learned of the magnetic compass by contacts with China through central Asia, while the sailors of Islam probably saw it being used in the south China Seas, acquiring further knowledge from the European mariners.

By 1300 AD two major improvements had been made to the original form of the magnetic compass. First, the compass itself was set in gimbal rings. These were a series of concentric rings of brass so pivoted that when the ship rolled or tossed the compass remained upright. Gimbal rings are known to have been used in Europe for supporting censers in churches from 1230 AD, although their use in central Asia and China was even older. There is some uncertainty, therefore, as to where gimbals were first used to steady navigational compasses. A surviving example of Arab gimbals, dated 1271 AD, in the British Museum, shows how rapidly their use spread in the Old World.

The second important innovation was the introduction of the wind rose or compass card marked with the four cardinal points and their subdivisions. The wind rose is said to have been invented by Flavio Gioia of Amalfi in about 1300 AD. Whether or not this is true, it was undoubtedly an Italian innovation. Before the introduction of the compass card, the needle could be used only as a rough check on the direction taken by the ship, and the mariner still had to rely largely upon other factors such as the positions of the Sun, Moon and stars. The wind rose, however, allowed the navigator to set the helmsman's course directly, and make it possible for him to keep a continuous check on that course. Furthermore it allowed the mariner to establish his position at sea when in sight of land by taking bearings of points along the coast, such as headlands, which could be plotted on his chart. It was the wind rose more than anything else that made possible the series of explorations set in motion by the Portuguese prince, Henry the Navigator (1394–1460), opening up the route to the Cape of Good Hope and the Far East, and leading ultimately to the discovery of the New World and the great voyages of discovery of the early 16th century.

Above right: a brass bound lodestone, used for magnetizing compass needles. The method is to stroke the needle in one direction only, using one end of the stone, taking it well away from the needle at the end of each stroke.

Right: folding pocket sundials, such as this one, were popular in the seventeenth century. A compass was necessary so that the sundial could be aligned properly, with its shadow stick or gnomon, in this case a string, pointing north-south.

MICHAEL HOLFORD LIBRARY/SCIENCE MUSEUM

MICHAEL HOLFORD LIBRARY/SCIENCE MUSEUM

The quadrant

Equipped with an astrolabe, the mediaeval scientists could make astronomical observations, tell the time, and undertake all the geometrical techniques of surveying, gunnery and navigation. These ornate instruments, which in practice were rarely used but kept for show, were combined analog computers, astrological predictors and angle measuring devices, with both sides covered with as many scales and symbols as could be managed. Their very comprehensiveness meant that many features were redundant for use in particular applications such as navigation or surveying.

A quadrant is a quarter of a circle, containing 90°, cut out of sheet metal. Essentially it is a much simplified version of the angle-measuring part of the astrolabe, and measures the angle between a line of sight set along one straight edge and a plumb line suspended from the apex. Such instruments were used by mediaeval astronomers to measure the altitudes, or angles above the horizon, of stars, but these early quadrants still frequently contained additional scales for other uses. By the sixteenth century, however, there were individual quadrants designed for specific uses—astronomy, surveying, gunnery, time telling and navigation, for example.

The earliest European description of a quadrant was written by Leonardo of Pisa about 1220. This was a surveying instrument some six inches (15 cm) in radius with a calibrated circumference and a shadow or geometric square to provide a direct solution to problems involving angles and sides of triangles. There was nothing in this quadrant not to be found on the reverse face of an astrolabe.

The earliest surviving European quadrant, *quadrans vetus*, is an *horary* quadrant: that is, an instrument for telling the time. The very name, the 'old quadrant', suggests a superseded instrument, and mediaeval manuscripts refer to a 'new quadrant', *quadrans novus*, designed by Prophiatus Judeus of Montpellier in the late thirteenth century. None of these horary quadrants had any great advantage over the astrolabe other than smaller overall size and cheapness of manufacture. During subsequent centuries many other horary quadrants were designed, which were basically altitude sundials, telling the time from observation of the altitude or elevation of the sun above the horizon.

During the sixteenth century the ancient science of astronomy was being radically changed. The great Danish astronomer Tycho Brahe stressed the vital importance of accurate quantitative observation of celestial phenomena, and himself refined instrumental and observational techniques. His momentous achievements were enshrined in his 6½ foot (2 m) radius astronomical mural quadrant of 1582—called mural because it was fixed to a wall. By virtue of improved read-out facilities Tycho was able to measure angles to 10 seconds of arc. Large fixed astronomical or mural quadrants built in to the structure of observatories became commonplace by the eighteenth century. By that time improved methods of edge division and the use of refined techniques of reading-out small fractions of a degree based on Pierre Vernier's vernier gauge of 1631 and William Gascoigne's screw micrometer of about 1639 gave astronomers even greater accuracy of measurement.

The eighteenth century also saw a

RONAN PICTURE LIBRARY/ROYAL ASTRONOMICAL SOCIETY

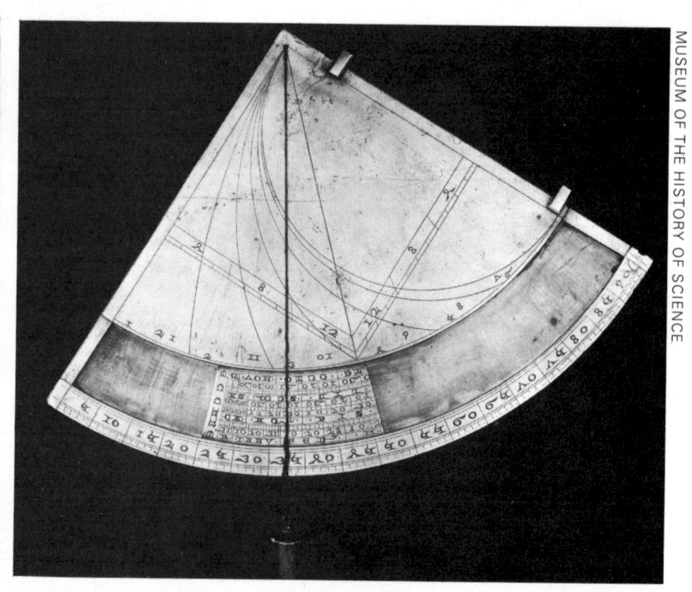

MUSEUM OF THE HISTORY OF SCIENCE

Below left: an illustration of 1508 showing the Muse of Astronomy giving the Greek scientist Ptolemy a lesson in the use of the quadrant to measure the Moon's altitude. This unlikely circumstance is made even less credible by the fact that the quadrant's plumb line, against which the angle is measured, seems to be missing.

Below right: a 'quadrans vetus' dating from about 1300. This was used for telling the time by measuring the sun's midday altitude; half the sliding cursor, here set for latitude 44°, is missing.

radical improvement in navigational technique. The navigational quadrant of late mediaeval times was no more than a simply divided arc with which the mariner could observe the noon elevation of the sun. This was the only reliable method of fixing the latitude of a ship, but reading the swinging plumb line on a vessel bobbing on the waves was practically impossible. Robert Hooke in 1670 and Sir Isaac Newton in that same decade both proposed the idea of a reflecting quadrant using the horizon as a reference. In 1731, quite independently and totally ignorant of these earlier forgotten ideas, John Hadley in London and Thomas Godfrey in Philadelphia both designed reflecting quadrants for the use of seamen. In the Hadley quadrant, as it was called throughout the eighteenth century, the image of the sun was observed after reflection through two mirrors.

The result of this arrangement was twofold. Firstly, because of the optical principles governing double reflection, the frame of the Hadley quadrant was not 90° but half that angle—an eighth of a circle—hence the instrument's other name, octant. The second and supreme feature of the reflecting quadrant was that by virtue of the two mirrors the navigator observed the image of the sun at noon as if it were on the horizon. All the problems of precise observation on the deck of a ship tossing in the waves were thus reduced from the practically insoluble to the achievable. This invention gave a tremendous impetus to scientific navigation.

Quite naturally there were many minor improvements, the most notable being that of Captain John Campbell about 1757. In Campbell's reflecting quadrant the size of the arc of the scale was increased to read to 120°, the instrument then being one sixth of a circle. This became known as the sextant, an invention that still remains an indispensible part of the equipment of every navigator.

Below: this 17th century quadrant for surveying carries a vernier scale between the two radial arms for more accurate measurement of the angles—a precision of 1 minute of arc is possible.

Bottom: Hadley's quadrant, from a book of 1788. The small diagrams show how turning a mirror through a certain angle reflects its image through twice that angle. Fig 19 shows the principle of use—the horizon is viewed past mirror F which reflects the sun via movable mirror G.

THE SCIENCE MUSEUM

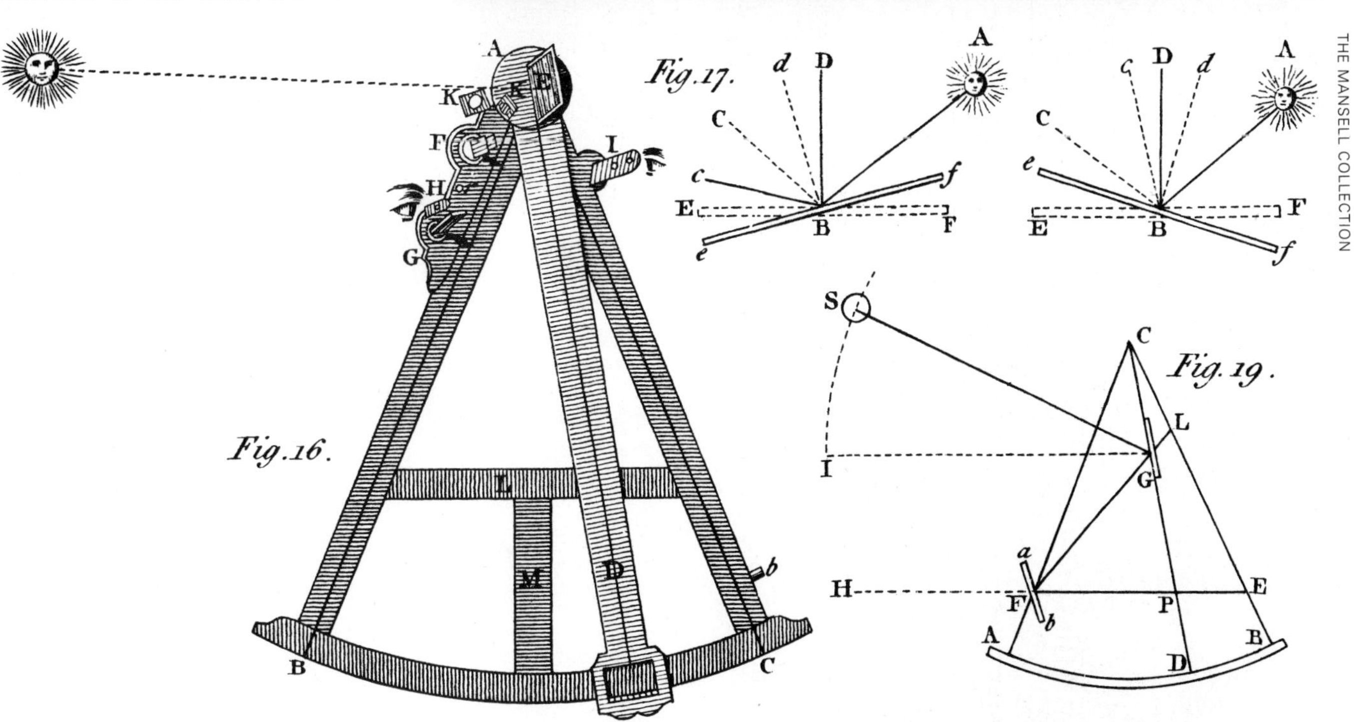

THE MANSELL COLLECTION

Knitting machinery

The origin of knitting is uncertain though it is most likely that it began in the Middle East. The earliest known knitted remnant has been dated to 256 AD, and further specimens have been found in the tombs of Egyptian kings from the 5th century AD.

The basis for all knitting is the loop, unlike weaving, which consists of threads interwoven at right angles. The weaving process is simpler to understand, but requires a framework of some size. Knitting, however, can be carried out on a smaller scale but requires more complex stitches.

Before the sixteenth century all knitting was done using a pair of rods—such as domestic knitting needles—or on a *peg frame*, which could be flat or circular with a number of evenly spaced pegs. The simplest example, and perhaps the most well known, is the child's *French knitting* made on four pins around the central hole of a cotton reel. The peg frame, with a number of pins in line, rather than in a circular pattern, is the basis of knitting machines.

Each peg has a small *cross piece* on top, to hold the loops of wool in place. The principle is that, initially, the yarn is looped around each cross piece in the line, and then, reversing the direction, the yarn is held across the front of each loop in turn. The earlier loop is slipped off its peg (*casting off*), over the yarn which then forms the next loop on the cross piece.

The process continues for each peg, until a new line of loops has been formed. The direction is reversed again, a new line made, and so on, the resultant knitting travelling down the pegs. If the yarn is laid behind the loops, *purl* knitting results. This type of frame has long been used for the knitting of tapestries and carpets, but the man who turned the principle into a practical knitting machine was the Reverend William Lee.

There are several stories as to how Lee, an English country parson of the sixteenth century, came to invent the knitting machine. The stories tend to involve a young woman he wanted to marry whose preoccupation with hand knitting made him resolve to invent a machine to do the work instead. Alternative, but similar, tales tell how he invented the machine to save his wife work, after he had been expelled by

THE SCIENCE MUSEUM

LEICESTER MUSEUM/MIKE ST MAUR SHEIL

Left: William Lee's original machine is now no longer in existence, but this 1770 version is very similar. Foot treadles were used to operate the sinkers.

Above: detail of work on an 1820 hand frame machine. Here the yarn is being laid across the tops of the needles. After this, the sinkers are lowered, forming loops, then the presser bar is lowered to trap them.

Right: bearded needle and three types of latch needle, which superseded it.

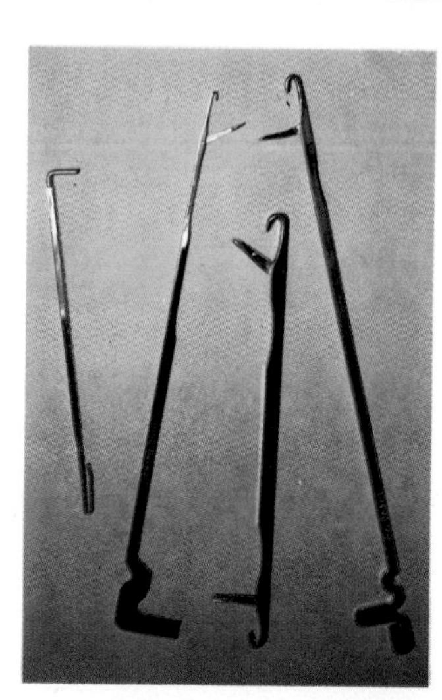

his university for marrying below his station.

How near the truth these are is debatable. We do know, however, that Lee was born in Calverton, Nottinghamshire, in about 1564. In 1586 he returned from Cambridge University to find his parishioners supplementing their income by hand knitting. For whatever reason, Lee decided to try to mechanize this activity, and the final design of the machines has much more relationship to peg frames than to any domestic needle knitting.

Much of Lee's inventiveness must have been taken up in devising the needle to replace the pegs, so that the loops would be held in place. He finally came up with the *bearded*, or *spring beard* needle, the principle of which is still in use today.

A line of needles is used, each needle replacing a peg of the frame. Each has a springy hook, whose long tapering end is called the *beard*, which can be pressed against the body or *shank* of the needle. This makes it possible to retain one loop of yarn while the previous loop is cast off over it. The formation of loops is carried out by two alternate series of shaped bars called *sinkers*, operated by foot treadles, and the pressing of the beard down to retain the loops is done by a *presser bar*. The early machines required a certain amount of manual work, but the operation was nevertheless much faster than hand knitting.

This type of machine was in use for several hundred years. It was not accepted at first, particularly in Lee's time, because it was feared that it might cause unemployment and also for the much better reason that the quality of the early machines left something to be desired. Later machines were capable of finer work, however, and their use spread, first in a cottage industry system and then in full factory use around the time of the Industrial Revolution. Wood framed machines were replaced by iron ones, and powered versions known as *rotary frames* were introduced.

For 260 years the bearded needle reigned supreme in machine knitting despite other machine improvements. In 1849, however, a Leicester hosier named Matthew Townsend invented the *latch needle*. The hook of this needed no presser bar to operate it, since it was hinged. Lee's bearded needle is still widely used, though, as it can be made much finer and can give a closer stitch.

LEICESTER MUSEUM/MIKE ST MAUR SHEIL

Above: a circular knitting machine made in America during the 1920s.

Below: detail of the yarn feed. This uses latch needles. The knitting is continuous and forms a tube 27 inches (69 cm) in diameter. This can then be used as part of a garment, or slit open and stitched.

The barometer

The basic principle of the barometer is to use atmospheric pressure to balance a vertical column of mercury sealed at the top. The height of the mercury indicates the atmospheric pressure. The phenomenon of atmospheric pressure, however, was not always known, nor, indeed, that of a vacuum. The Greek philosopher, Aristotle, had taught that a vacuum was a logical and physical impossibility, but he believed, correctly, that air, the atmosphere surrounding the Earth, had weight. Almost two thousand years later, Renaissance scientists maintained the impossibility of a vacuum. They were, however, firmly convinced that air had no weight. The early history of the barometer is closely related to the initial investigations of the phenomenon of the vacuum.

In 1638 the famous Italian scientist Galileo Galilei published his last book, *Two New Sciences.* While Galileo maintained that air had no weight, and thus could not exert a pressure, he did believe in the vacuum. His comments on this subject stimulated others such as Otto von Guericke to investigate the properties of a vacuum.

About 1640 a physicist, Gasparo Berti, fastened a long lead tube to the wall of his house in Rome. The foot of the tube was fitted with a stop-cock and set in a vessel partially filled with water. A glass receiver was fastened at the top of the tube. Tube and receiver were filled with water and securely sealed. Then the stop-cock was opened. It was observed that some, but not all, of the water flowed out into the vessel. A column of water, subsequently measured as 18 cubits (32 feet) in height, was left in the tube and above it was an apparently empty space. We know that there was a vacuum in that empty space and that the column of water was supported by the weight of the atmosphere pressing down on the water in the lower vessel. In effect Berti had made a water barometer; however neither he nor his colleagues understood fully the experiment he had performed.

Writing from Florence in 1644, Evangelista Torricelli described a similar experiment. A glass tube two ells (about 36 inches) long was filled with mercury. The open end was closed with the thumb and the tube inverted and held upright. After the open end had been placed beneath the surface of some mercury in a trough, the thumb was removed. Then the level of the mercury in the tube dropped, leaving a space of about four inches. Torricelli's assistant Vincenzo Viviani probably performed the actual experiment and may have had the clever idea of using the dense liquid-metal mercury, so reducing the size of the apparatus. Torricelli, however, unlike his contemporaries, had a total comprehension of the physical principles involved in this experiment. Furthermore, Torricelli envisaged the possibility of using the variation in the height of the mercury in the tube above that in the trough to find out when air was 'heavier or lighter'. Here was the idea of an instrument to measure the pressure of the atmosphere; though Torricelli never proceeded to make an instrument, and it was some years before Torricelli's experiment became formalized into an instrument.

In 1647, Blaise Pascal, after several experiments, convinced people of the correctness of Torricelli's ideas. In 1648 his brother-in-law took a barometer up the Puy de Dôme mountain in the Auvergne and found that, as he climbed, the height of the mercury column

Berti's atmospheric experiment. At the time, its significance was not realized. The purpose of the bell M was to discover whether or not sound travels in a vacuum.

decreased. Aristotle's teaching of nature's abhorrence of a vacuum was disproved; Galileo's ideas of a vacuum were affirmed. The pressure of the atmosphere fell with increasing altitude.

In 1660 in his laboratory in Oxford, Robert Boyle set up Torricelli's experiment inside the receiver of an air-pump. As air was evacuated from the receiver, so the mercury sank. Incidentally it was Boyle who coined the word, 'barometer', which first appeared in print in 1665.

Almost as soon as the barometer had become widely accepted in the mid-17th century, scientists began to investigate a possible relationship between fluctuations in the height of the mercury and changes in the weather. Attempts were therefore made to design a more portable instrument and to increase the variation due to pressure, which is normally only two to three inches in the European environment. Pascal and Boyle independently suggested that portability could be achieved by dispensing with the separate mercury cistern and bending the bottom of the barometric tube through 180°, so creating a siphon barometer. In 1664 Robert Hooke published a design for a wheel barometer in which the movement of the mercury was magnified by a mechanical linkage. In about 1680 Sir Samuel Moreland suggested a diagonal barometer—bending the tube at about 27 inches above the cistern so that the last few inches of vertical height corresponded to a greater length of tube. In the 19th century the rise of meteorological science led to radical improvements in detailed design features and the barometer became a reliable and accurate scientific instrument.

Below: one of Torricelli's experiments. The bladder, filled with a small amount of air, expanded as the barometer was lifted in the mercury. A vacuum was formed at the top, and the excess pressure inside the bladder expanded it.

RONAN PICTURE LIBRARY/E P GOLDSCHMIDT

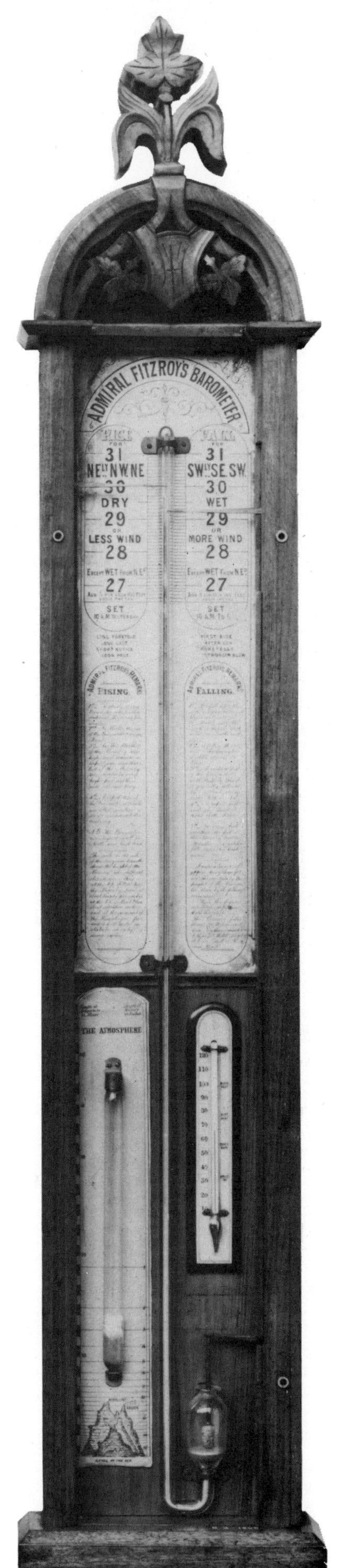

THE SCIENCE MUSEUM

Right: this popular Victorian barometer gives detailed instructions for making home forecasts:
Rising.
1st—a steady rising barometer which when continued shows very fine weather.
2nd—in winter the rise of the barometer presages frost.
3rd—in wet weather if the mercury rises high and remains so expect fine weather, but if the mercury rises suddenly very high fine weather will not last long.
4th—a rapid rise of the barometer indicates unsettled weather, a slow movement the contrary.
NB The barometer rises highest of all for north and east wind.
Falling.
1st—if a fall takes place with a rising thermometer wind and rain may be expected from the south eastward, southward, or south westward.
2nd—a fall with a low thermometer foretells snow or rain.
3rd—a sudden fall of the barometer with westerly winds is generally followed by a violent storm from NW or NE.
4th—a rapid fall indicates wind or wind with rain.
5th—in very hot weather the fall of the mercury denotes thunder, or a sudden fall indicates high wind.

Indications of approaching changes are shown less by the height of the barometer than by its falling or rising.

Thus the figures are of more importance than the words.

The mercury falls lowest for wind and rain together, next to that for wind except it be an east or north east wind.

Firing mechanisms

The principle of the *flintlock* was simple and in comparison with earlier firing mechanisms (the *matchlock* and the *wheel-lock*) more reliable. When a piece of flint was struck against a hardened steel plate, it produced a shower of sparks which fell into the priming pan below, igniting the gunpowder prime. The flash passed through a small vent and ignited the main charge in the barrel. Various designs were developed for igniting all types of firearms from pocket pistols to large cannon.

The earliest form of flintlock, first recorded in Sweden in 1539, was used throughout northern Europe for about 200 years, and nowadays is referred to as the 'Baltic lock'. Mechanically it was of the *snaphance* type, that is with a separate priming pan cover and steel, and a horizontally acting *sear*—the mechanism for transmitting the trigger action to the *cock* which held the *flint*. Another snaphance mechanism appeared in southern Europe, in Florence, in 1547, but there is reasonable evidence to suggest that the original idea for the flintlock originated in southern Germany.

The snaphance (also correctly spelt snaphaunce) was in use in France sometime after 1550, and first recorded in England in 1580, but was not recorded in Italy until the early 17th century. The Italians, however, continued to use this type of mechanism when the remainder of Europe, with the conspicuous exception of Spain, had changed to the French flintlock.

The Spanish adopted a mechanism known as the *miquelet* which combined the earlier horizontally acting sear with the united steel and pan cover of the true flintlock.

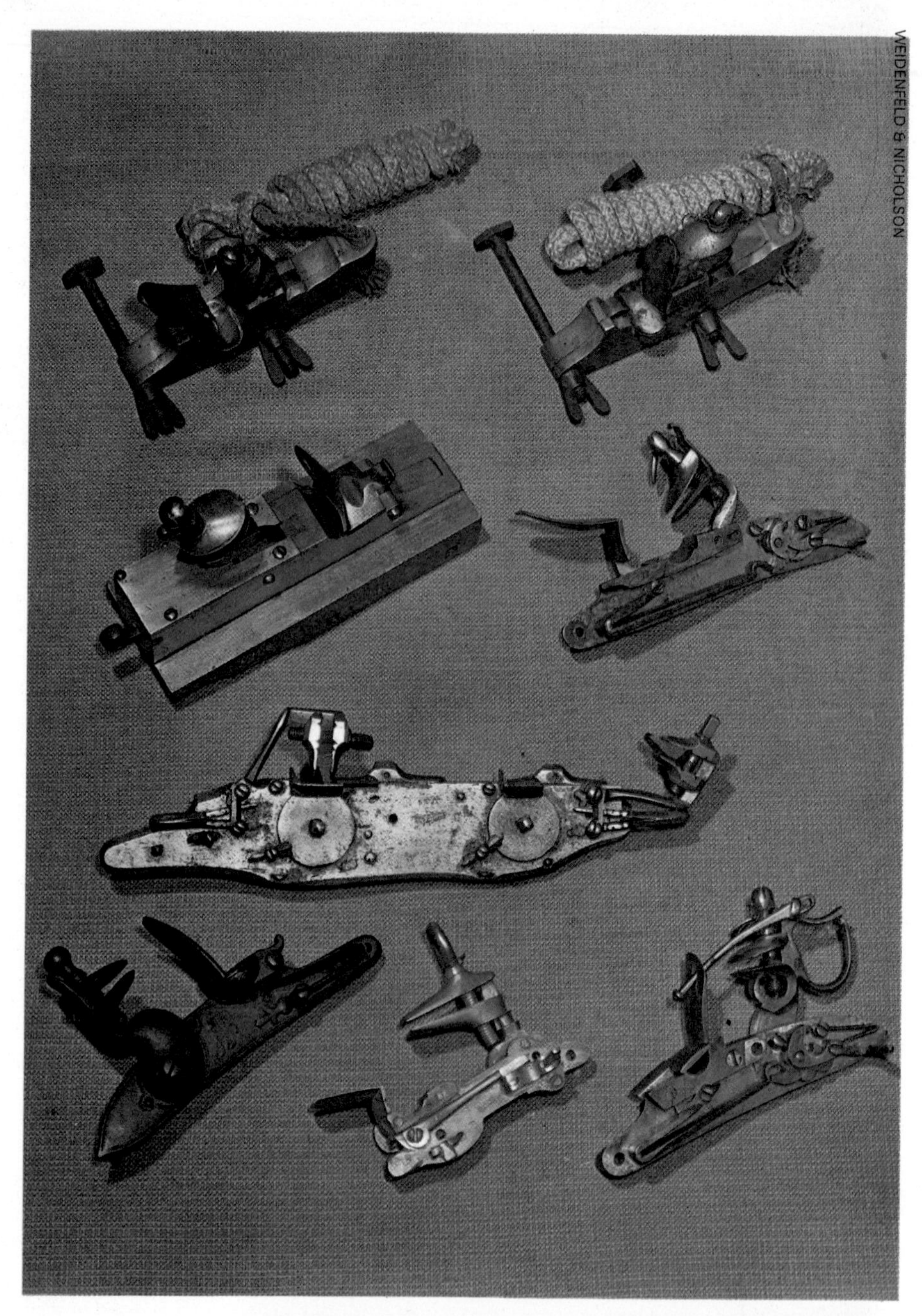

WEIDENFELD & NICHOLSON

Top: various types of firing mechanisms. From left to right beginning at the top: the first three flintlocks are similar and were bolted to the breech of muzzle-loaded cannon, replacing the older linstock; the lanyard was used for firing—an inside view of a typical flintlock, used on an Austrian musket of 1822—a curious double wheellock, forerunner of the flintlock—a Portuguese copy of a British Land Pattern lock, typical of those used on most 18th century weapons—a Spanish miquelet—a French lock designed to be fired with a cover over it for protection in wet weather.

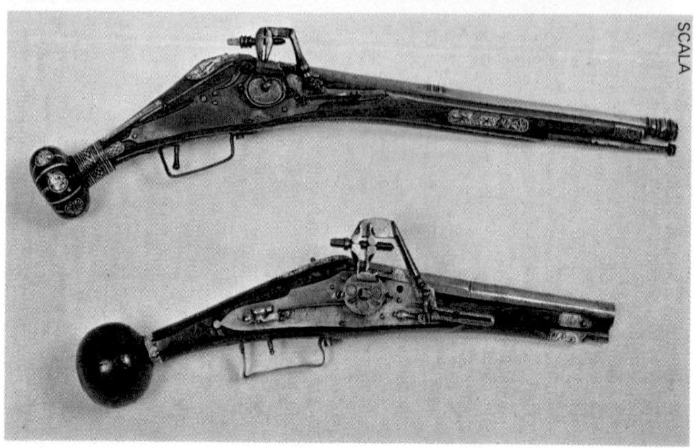

SCALA

Right: 16th century Venetian firearms with wheellocks. A spring, under tension, was released by pressing the trigger. This caused a wheel with a serrated edge to revolve rapidly against a piece of pyrites, producing sparks which ignited the priming powder.

The 'true' or French flintlock is now generally accepted as having been devised by a clock- and gunmaker of Lisieux (a provincial capital some 25 miles due south of Le Havre), named Marin le Bourgeoys, during the first decade of the 17th century. Its features included a steel and pan cover combined to form one L-shaped piece, and a vertically acting sear which engaged notches cut into the edge of a 'tumbler' thereby controlling the action of the cock which held the flint in its jaws. The L-shaped piece is variously called a 'hammer', 'battery', 'frizzen' or 'steel', the two former being contemporary terms and the two latter more modern in origin.

The general adoption of this form of flintlock owes a great deal to the expansion of France during the 17th century; as with most social arts, firearms design increasingly followed the lead set by the French with the result that the Low Countries and Germany, the great arms producing areas of the time, rapidly adopted the improved mechanism, while gunmakers to the nobility and gentry in all countries aped the French designs in both mechanism and decoration for their 'arms de luxe', or prestige weapons. And it was the French, in the 1670s, who led the change to flintlock weapons for military use. The only areas to successfully resist this trend were Spain and Italy, but by the second quarter of the 18th century even these countries had come round to the French patterns in all but detail and only the miquelet managed to survive as a distinct variation until the 19th century.

The flintlock was the culmination of a long series of designs utilizing a small amount of external priming powder, which was ignited by a mechanism either holding or producing a spark, and requiring an opening at the breech of the barrel through which the flash from the external explosion must pass to fire the main charge. All these designs were subject to unreliable performance in wet or windy weather, and could not be left either loaded or primed for any length of time without the probability of failing to fire.

When, therefore, the *percussion lock* was developed and perfected between 1805 and 1825, it rapidly replaced the flintlock since it did away with a breech which was open to the weather, and confined the priming to a relatively waterproof encapsulation. It also greatly simplified the construction of the mechanism needed to ignite the priming.

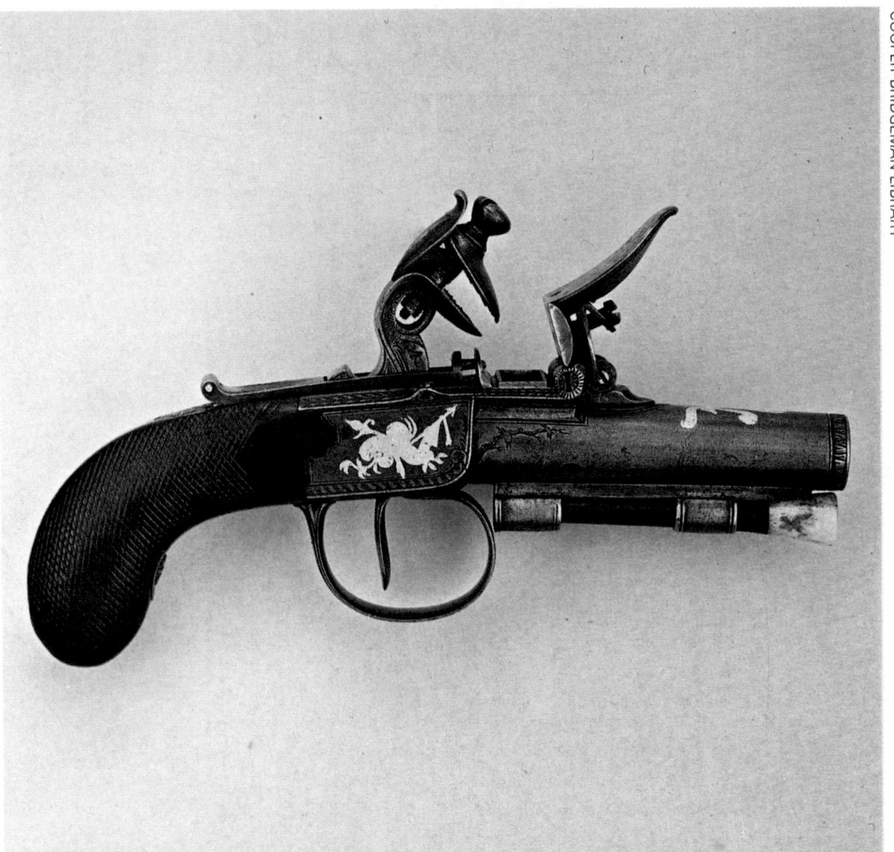

COOPER-BRIDGEMAN LIBRARY

PHOTRI

Top: a pocket pistol made by the famous gunsmith Durs Egg, c. 1790, for the Prince of Wales. It has a typical 18th century flintlock with L-shaped steel and pan cover.

Bottom: in the 19th century, the percussion lock replaced the flintlock. The flash came when a cap, covering a hole leading to the barrel, was struck by the hammer.

The microscope

Single lenses were being used to aid vision in spectacles and magnifying glasses by 1300. Such devices, however, have limited magnification and it was 300 years before two such lenses were used in combination to give the compound microscope. This instrument was probably invented by Hans and Zacharias Janssen of Middelburg in Holland about 1600. Their microscope was a tube about 18 inches long and 2 inches in diameter, with a single convex lens at each end: it had no stand, but was held in the hand and could only be used with reflected light.

Galileo helped to make the instrument popular in the first half of the 17th century, but it was not until the publication in 1665 of Robert Hooke's book *Micrographia* that men of science began to be seriously interested in the possibilities of the microscope. Hooke's microscope was better supported than earlier models, and could look at objects which were transparent as well as opaque. Similar microscopes were made for many years afterwards, by famous craftsmen such as Marshall, and it became a matter of status to possess one. Unfortunately, although progress had been made in refining details of the stand, the lenses remained very crude, so the magnification was low and the range of subjects limited.

It may have been the very poor results given by the compound microscope which persuaded one of the most famous names in the history of microscopy, Leeuwenhoek, to use very high-powered simple microscopes in his researches. Before he died in 1723, Leeuwenhoek spent over 50 years as an amateur of the microscope, making his own instruments and discovering bacteria and many important plants and animals.

During the 18th century the instrument was little better than a plaything, unused by serious men of science, and even distrusted by them. Some developments to the stand took place, one of the first being the design of a tripod microscope by Culpeper. From about 1730 this displaced the Marshall type, even though its performance optically was no better. Cuff introduced a much improved stand in 1744, and this is the forerunner of that found on a modern microscope. Two other 18th century designers require mention—Martin and Adams. Both made instruments and also wrote books, and their works had important results in renewing interest in the microscope. Martin's 'New Universal' microscope of 1770 was well designed and very practical. The silver microscope made by Adams for King George III was not at all practical, but certainly impressive. He also made much simpler and more useful microscopes.

Below: bearing no resemblance to today's microscopes, Leeuwenhoek's, c 1673, consisted of a double-convex lens mounted between two brass plates. The long screw positioned the specimen on the point of a rod mounted on a brass block, which could be moved against the plate to focus the specimen (viewed from the other side) by adjusting the short screw.

Right: a microscope and stand made in 1746 by George Adams, the optician and instrument maker, who wrote 'Micrographia Illustrata'.

RONAN PICTURE LIBRARY

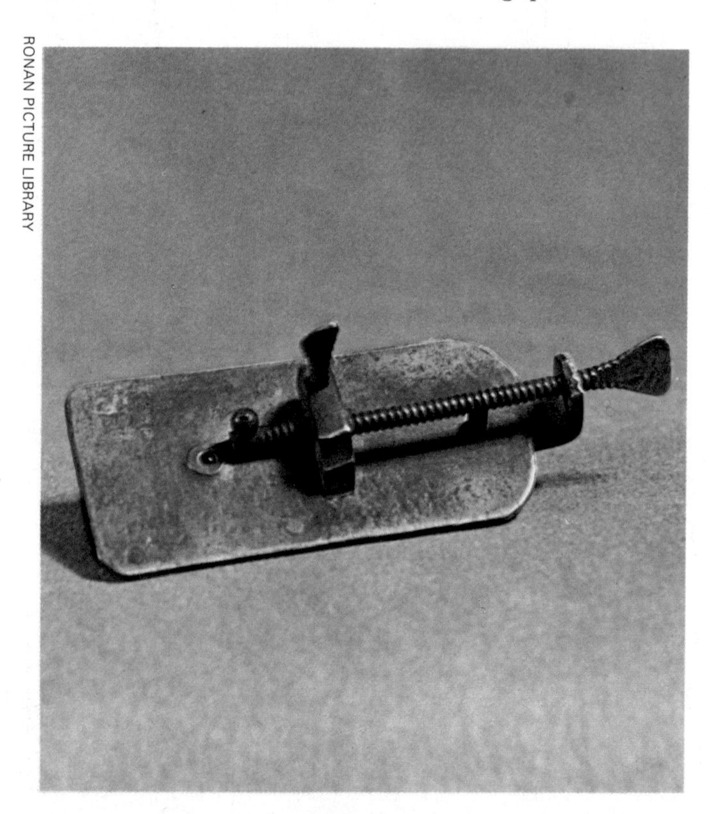

THE SCIENCE MUSEUM

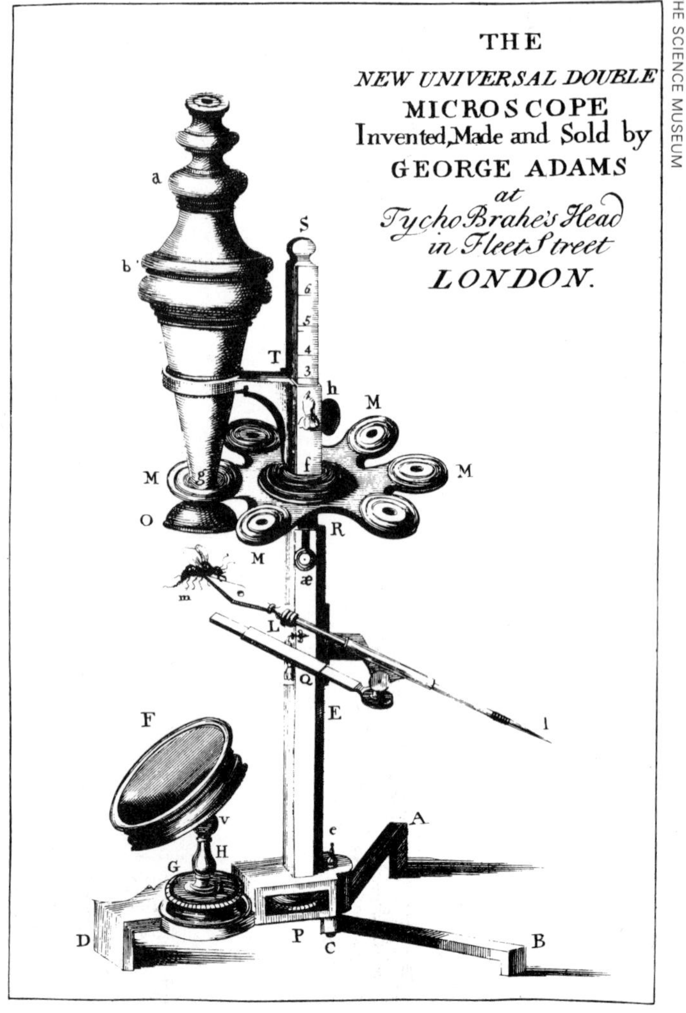

The lenses on all microscopes suffered from two severe defects—chromatic and spherical aberration, apart from giving only low magnifications. So severe were the aberrations that some scientists believed that matter was made up of minute globules, whereas in fact the globules were actually optical phenomena resulting from poor lenses. By 1830, however, the microscope objective had been improved to an enormous extent, largely through the work of J J Lister, a wine merchant and father of the famous Lord Lister. He invented the *achromatic* and *aplanatic* objective which corrected the two aberrations, and allowed small detail to be seen clearly. Following this breakthrough, opticians vied with each other to produce better lenses and the stands to go with them. In England three makers were supreme—Powell & Lealand, Ross, and Smith Beck & Beck. In Europe Zeiss and Nachet were well known, while in the USA Tolles and others became notable. The use of the microscope became standard in the laboratory.

Zeiss employed and later took into partnership Ernst Abbe, who studied the optical theory of the microscope in detail. He applied this theory to the making of lenses, and by 1870 all Zeiss lenses were the fruit of carefully-followed mathematical design, rather than the chance of the skill of an individual craftsman. Two other important advances resulted from Abbe's work—oil-immersion objectives and *apochromatic* (having the greatest correction for spherical and chromatic aberrations) lenses. The English microscopist, Stephenson, suggested to Abbe that improvements should result from immersing the front of a special objective in a liquid of the same refractive index as the glass, and this proved to be so when it was tried in 1878. By 1882, using glass made by Schott, Abbe succeeded in making the most perfectly-corrected lenses the world had ever known—the apochromatic series. This was the perfection of the work started by the older Lister over 50 years before, and the history of the microscope since that time has been one of simplifying the controls and developing phase-contrast and interference systems. Its resolution and magnification cannot be bettered.

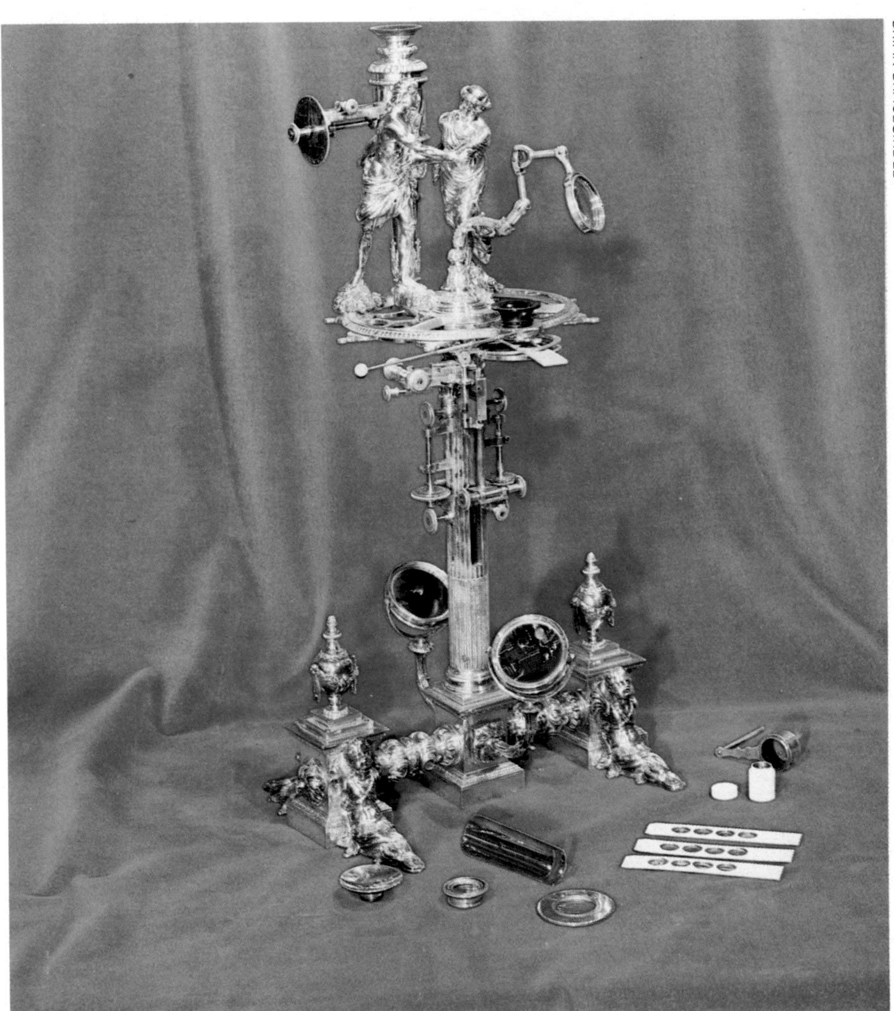

BRIAN BRACEGIRDLE

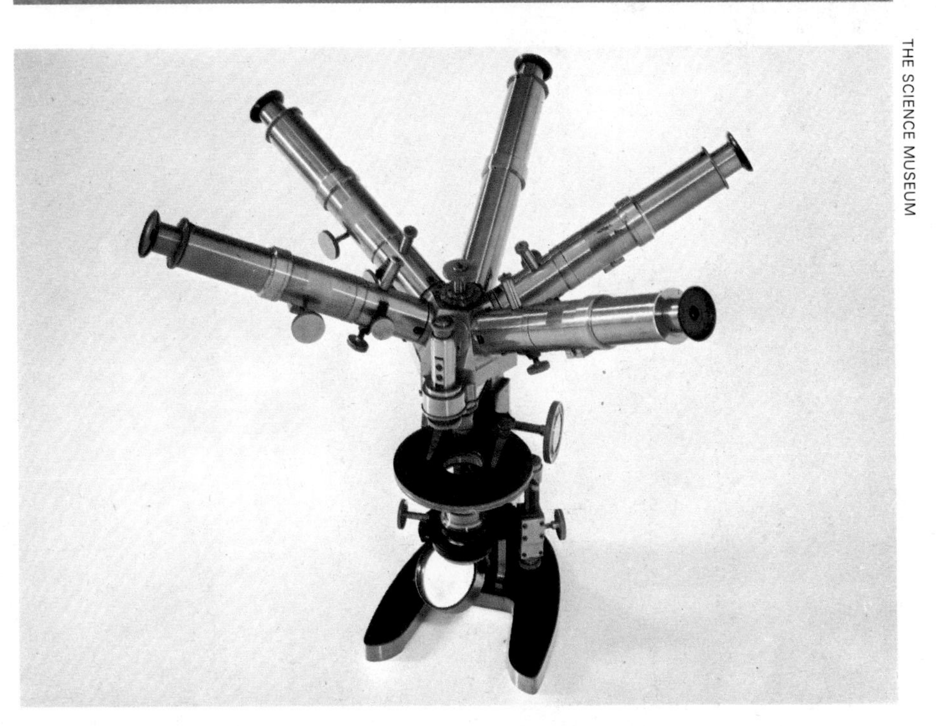

THE SCIENCE MUSEUM

Top: one of the most elaborately decorated instruments was this silver microscope made for George III by Adams. It was not very practical, as it was about 29 inches high.

Bottom: an unusual late 19th century model made for the Swiss Professor Thury for class work. It has 5 eyepieces mounted on a central box containing a 90° prism which can be rotated to direct light from the single objective into one of the eyepieces, after first being focused with the central eyepiece.

The thermometer

RCNAN PICTURE LIBRARY

Santorio's air thermoscope, with the upper bulb in the patient's mouth.

Even before such a thing as a thermometer was invented, doctors were using subjective scales of temperature. In the 2nd century AD the Greek physician Galen of Pergamon proposed that the sensations 'hot' and 'cold' could be measured by reference to a scale of four degrees of heat and cold numbered above and below a neutral point, and Galen's teaching formed the basis of medical practice until the 17th century.

The writings of Hero of Alexandria, who probably lived in the 1st century AD, were well known to Italian scholars following the publication in 1575 of an edition of his manuscripts. Among the strange devices described by Hero was a 'fountain that drips in the sun'. Eventually the Italian scientists who investigated this device were able to propose an explanation—expansion due to heat. By the end of the 16th century they had developed the *air-thermoscope*, the immediate precursor of the thermometer. It illustrates qualitatively the effect of heat by the expansion of air.

The simple air-thermoscope consisted of a glass flask with a long thin neck. The flask was warmed, expanding and partially expelling the air. Then the inverted flask was held with the mouth of the neck under water with the result that when the heat was removed, the contracting air drew water up the neck. Subsequent alternations in the height of the water in the neck of the thermoscope

Below left: a copy of Galileo's air thermoscope of 1592. These devices used air as the thermometric substance: its pressure variations with temperature altered the height of the red alcohol.

Below right: a copy of an early sealed alcohol in glass thermometer as made in Florence. The helical scale made far more sensitive readings possible; the divisions were marked by small black glass beads.

MICHAEL HOLFORD LIBRARY

SCIENCE MUSEUM/MICHAEL HOLFORD LIBRARY

indicated temperature changes. In 1611 Bartolomeo Telioux of Rome illustrated a thermoscope with a scale, but his description was second-hand and lacked comprehension of the physical principles involved. An Italian doctor, Santorio Santorre, has the best claim to the accolade 'inventor of the thermometer'. By 1612 Santorio had developed the air-thermoscope with a scale of degrees attached to provide a quantitative measure of temperature.

By the 1650s the air-thermometer was well known: in 1660 Otto von Guericke erected an impressive air-thermometer on the wall of his house in Magdeburg. By this time, however, the reliability of the air-thermometer was being seriously questioned. In 1644 Evangelista Torricelli had discovered the variability of the pressure of the air, and during the 1660s it was realized that the air-thermometer reacts to changes in pressure as well as the change in temperature. A solution to this problem already existed: this was the sealed liquid-in-glass thermometer, which had been invented in or before 1654 by the Grand Duke of Tuscany, Ferdinand II. This instrument is not affected by changes in atmospheric pressure, since any such changes act equally all over the device.

The sealed liquid-in-glass thermometer became widely known following the publication of the proceedings of the Accademia del Cimento of Florence in 1666. The Academy's glass-blower made 50° and 100° thermometers that look similar to the instrument we still use today, but with scales of degrees that differed from type to type, though two similar thermometers would agree. A 420° thermometer with a helical tube provided a marvellous demonstration of the expansion of the alcohol used as the thermometric liquid, but it was not really an instrument. In contrast the 50° Florentine thermometer provided a reliable standard. This was a remarkable achievement, for these thermometers were made before there was an understanding of the need for careful calibration using fixed points to define a universally comparable temperature scale.

During the 1660s Robert Hooke, Curator of Experiments to the Royal Society of London, tried to construct thermometers that would agree one with another without following the Florentine scheme of absolute physical similarity. Hooke adopted a single fixed point, the temperature at which water begins to freeze. Not until the 18th century were there successful attempts to produce comparable thermometers that could be made from first principles by any glass-blower. Notwithstanding this goal of a rational and universal scale of temperature, the 18th century saw a considerable number of thermometer scales: in 1778 a survey listed no less than 27. Proposals for using blood heat, the melting point of butter, or the temperature of the cellars in the Observatory at Paris as fixed points were discarded in time. By the 1740s there was growing agreement that the freezing and boiling points of water were the best datum points for thermometry.

In 1717 an instrument maker in Amsterdam, D G Fahrenheit, brought thermometers filled with mercury into commercial production. The temperature scale associated with Fahrenheit's name defines the freezing point of water as 32° and its boiling point as 212°. The centigrade scale has 0° at freezing and 100° at boiling point. Incidently, the International agreement of 1948 that this scale be named 'Celsius', after the Swedish astronomer Anders Celsius, perpetuates a historical error—Celsius' scale went the other way from our own, with freezing point at 100°.

Below left: two views of Otto von Guericke's air thermometer, the left hand one having its cover and scale removed to show the mechanism.

Below right: the centre illustration shows a 1764 thermometer operated by the expansion of metal rods.

THE SCIENCE MUSEUM

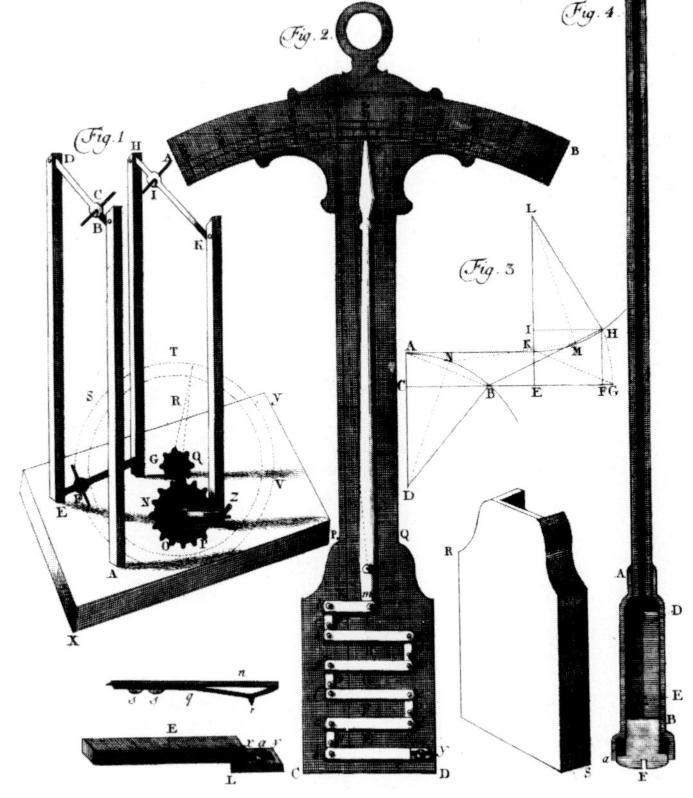

The telescope

The invention of the telescope is still something of a mystery, and even the discovery of its optical principles is a somewhat open question. Nevertheless, modern research makes it seem likely that the basic optics of instruments for 'seeing at a distance' were known at least as early as the thirteenth century, and it is virtually certain that telescopes were being used by a few scientifically minded men in England by the late 1570s. Even so the first claim for a patent only came decades later, in 1608, and then from the Netherlands, not from England, when Hans Lippershey constructed a working telescope to substantiate his claim. The instruments introduced there were crude by modern standards, but they stimulated a spate of telescope making which, a year later, led Galileo in Italy to begin constructing 'eye-glasses' of his own design. In January 1610 he had made an instrument that magnified thirty times, and the epoch-making results he obtained in studying the heavens stimulated development of the telescope all over Europe.

These early telescopes were invariably refractors, with a front lens to collect and focus the light, and an eyepiece at the rear for examining the image. They all suffered from two serious defects—spherical and chromatic aberration. Spherical aberration arises because lenses with surfaces which are parts of spheres (as all lenses then were) bring light passing through close to the centre to a different focus from light passing through close to the lens' edge. The result is that the image is only sharp at the centre or the edge, but never at both. Galileo attempted to overcome this by using a mask over the front lens, so preventing light from the edge ever reaching the eyepiece, but by the mid-17th century others overcame the defect by building telescopes of great focal length—a 60 foot (18 m) instrument being nothing uncommon. Chromatic aberration, giving a coloured fringe around objects, results from the dispersive effect of a lens, which brings light of different colours to focus at different points. For a time there appeared no solution to this defect: some tried to make lenses of a combination of materials whose powers of dispersion were different, such as glass and water, but not until the 1730s was the problem solved. Then a lawyer, Chester Moor Hall,

Right: an early English telescope, made about 1680. Like most other terrestrial telescopes, this one has draw tubes to shorten it when not in use, from which comes the word 'telescoping'. The word itself is Greek for 'far seeing'.

Below: Galileo's telescopes had a limited field of view, but an improved system, designed by Kepler, was made by Scheiner. He projected the Sun's image to study its rotation and sunspots.

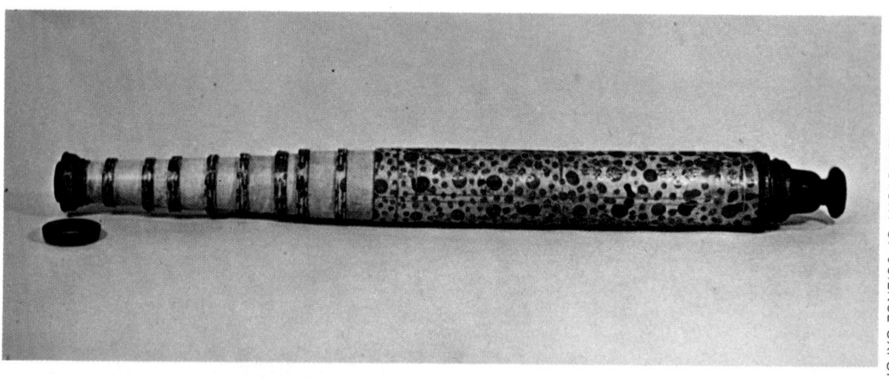

RONAN PICTURE LIBRARY/ROYAL ASTRONOMICAL SOCIETY

designed a double lens, with one component of the usual crown glass and the other of denser flint glass. The manufacture of such *achromatic* telescopes was taken up by many opticians, and since the double lens also reduced spherical aberration, there was no longer a need for refractors to be of such inconvenient lengths.

Although the earliest manufactured telescopes were refractors, it had long been realized that a concave mirror could be used instead of a lens to bring the light to a focus for examination by an eyepiece. It may be that some reflecting telescopes were constructed by experimenters in the 16th century, or even earlier, but the first practical reflector was designed in 1663 by James Gregory. Unfortunately its manufacture baffled opticians of the time and it was Isaac Newton who, in 1668, constructed the first wholly successful reflector as a solution to his attempts to overcome spherical and chromatic aberration. His design, using a small flat mirror to bring light from the main mirror to the eyepiece, was comparatively simple to construct. However, techniques developed so much in the next century that by the 1750s James Short was able to manufacture Gregorian type telescopes of fine quality, and soon opticians found that they could also construct the equally complex optics of the Frenchman Cassegrain's reflector.

The achromatic refractor, with its small aperture, held its own for astronomical observations where precision measurement was required, and for terrestrial use by mariners and surveyors. Yet space can only be probed successfully with large apertures, which is why, in the late 1770s, William Herschel turned to reflectors. Teaching himself to fashion optical parts, he built a series of reflectors of successively increasing size. His small telescopes had apertures of six to nine inches (15 to 23 cm) and were far superior to anything constructed before, but his crowning achievements were two instruments, one of 20 foot (6 m) focal length and having an 18 inch (46 cm) mirror, and the other a giant of 40 feet (12 m) focal length with a 48 inch (1.2 m) mirror—the largest telescope in the world when it was completed in 1789. No refractor could be made as large, and Herschel's success in observing dim and distant objects with reflectors made their advantages clear to astronomers. His huge 40 foot telescope was the direct precursor of the giant telescopes in today's observatories.

Below: even after the invention of the reflecting telescope, long refractors of extraordinary proportions were used. This one, with a lens made by Campani. was used in Rome in the early 18th century. Its size has been somewhat exaggerated, and its suspension mast is not shown.

Right: an Irish peer, Lord Rosse, built this 6 foot (1.8 m) aperture reflecting telescope at his home in 1845. It was then the world's largest telescope, but was of limited use because of its size.

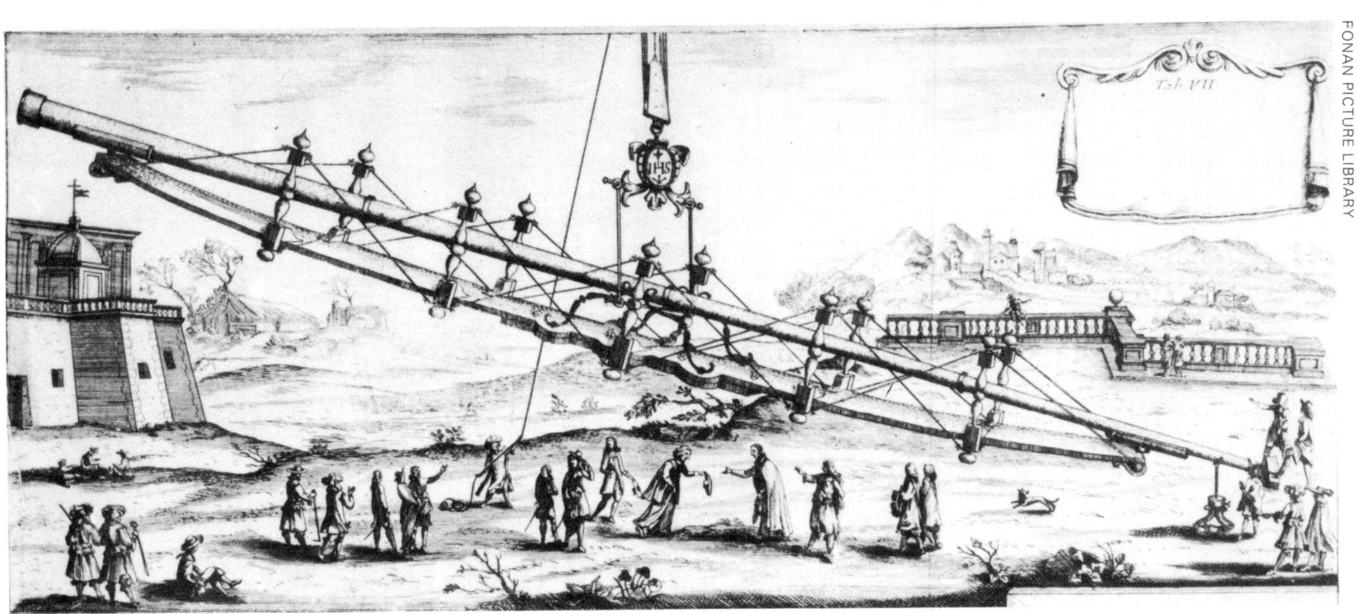

The pendulum

For three hundred years, it was impossible to make a clock which kept good time. The first mechanical clocks, made around the middle of the fourteenth century, were quite likely to be as much as an hour wrong, and they normally only had an hour hand. But from about 1660 onwards clocks which had previously been no more than general guides to the time could be given a quite simple modification which, almost literally overnight, turned them into accurate timepieces which were reliable enough to be given a minute hand. The invention which made this possible was the pendulum.

It is difficult to be precise about the 'invention' of something as simple as the pendulum: any swinging object has a regular beat, and can be used for timing purposes. There are reports that the Arabs used pendulums to time short events, by counting swings. What is more important is the recognition that the swings of any particular pendulum are indeed all of the same duration—that is, they are *isochronous*—and that this principle can be used to regulate a clock mechanism.

Galileo is widely regarded as being the first to realize this, and he probably was the first to give the notion any publicity. But the discovery in 1967 of more manuscripts by Leonardo da Vinci, which had previously been thought lost, show that Leonardo drew a diagram of a pendulum to regulate a clock mechanism as early as about 1495. There are suspicions that Galileo may have seen this, or heard about it, in 1637 when the manuscript changed hands.

Legend has it that Galileo first realized that a pendulum is isochronous in 1582, when he was 18. He is supposed to have timed the slow swings of a lamp hanging in the cathedral at Pisa, using his pulse. Following Galileo's discovery, astronomers and physicians used smaller pendulums to time events and patients' pulses. The limitation was that a small pendulum soon stops swinging, and has to be given a push every so often, which upsets the rhythm. In about 1641, Galileo designed a device which would

SCIENCE MUSEUM/MICHAEL HOLFORD LIBRARY

Left: a model of Galileo's escapement, reconstructed from his drawings, with a spring drive for demonstration purposes.

Below: an early engraving of Huygens' escapement. Flanges L are at an angle to each other: they strike teeth in turn.

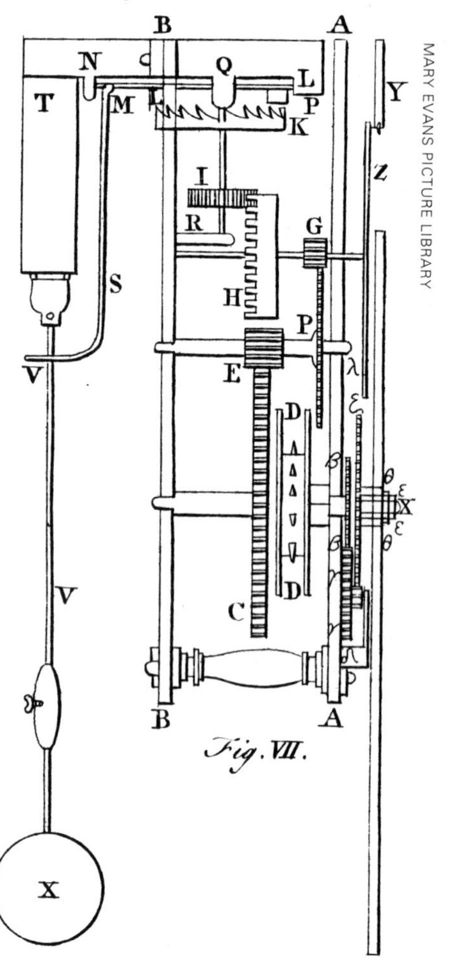

MARY EVANS PICTURE LIBRARY

give the pendulum a small push at the end of each swing, though he probably never made a working model.

All mechanical clocks need an *escapement*, which is the device that regulates the falling weight or the release of spring tension. In the early clocks the escapement was provided by a *foliot*—a weighted bar which takes some time to move when given an impulse by a tooth of a ratchet wheel. The foliot turns, but is stopped by a flange on its shaft that strikes another tooth of the ratchet wheel. The time taken for this to happen depends upon the inertia of the bar, but also on the freedom of its pivot. As this becomes clogged up, the clock runs slower and slower.

A pendulum escapement uses the driven ratchet wheel to keep the pendulum swinging, while the pendulum regulates the motion of the ratchet. Galileo's escapement was rather inaccurate, and it did not come into use. It was the Dutch scientist Christiaan Huygens who made the first working pendulum escapement in 1656, and made the invention known to the world.

Huygens, who did not know of Galileo's or Leonardo's escapements, devised a way of adapting the existing foliots to use a pendulum, by means of a simple gear train. He also made the first detailed study of the way a pendulum swings, and found that it is only isochronous for rather small swings. He worked out a way of altering the suspension of a pendulum so that larger swings were also isochronous, by means of a pair of 'cheeks' on either side of the pivot which the cable touched as it swung away from centre. These had the effect of slightly shortening the cable and made the pendulum completely isochronous.

Huygens' escapement was not ideal, but it had the advantage of being adaptable to existing clocks, and showed the way for further developments. These included the *anchor* and *deadbeat* escapements, which have rocking bars with a shaped tooth at either end, engaging in shaped teeth on the ratchet wheel. These do not interfere as much with the natural swing of the pendulum as did earlier escapements.

The introduction of the pendulum meant more than the facility to know what the time of day was: it enabled scientists to make measurements which involved the dimension of time far more accurately, the basis of the science of dynamics which led to a more complete understanding of the Universe.

Below left: original 1392 clock movement from Wells Cathedral, Somerset, adapted for a pendulum with an anchor escapement.

Below: this 1769 clock has a compound pendulum to compensate for changes in length with temperature.

SCIENCE MUSEUM

ROYAL OBSERVATORY GREENWICH / MICHAEL HOLFORD LIBRARY

Gunpowder

Gunpowder is one of the most important inventions in the history of mankind. Although its application has generally been for destructive purposes, its industrial and commercial uses have had significant and far-reaching effects, permitting a more economic exploitation of natural resources than otherwise would have been possible with the relatively crude machinery which was available before the nineteenth century.

Considering the impact that gunpowder has had upon the history of the world, it is amazing that its origins remain shrouded in uncertainty, despite recent intensive research. It is known, however, that a form of weak gunpowder was used by the Chinese early in the eleventh century, and that by the middle of the thirteenth century it was being used as a propellant. Gunpowder became known to Europeans through Arab scholars, and it is certain that Roger Bacon knew of the substance and its explosive effect in the 1290s.

By 1320 we have firm evidence of the use of gunpowder in Europe as a propellant charge for cannon and other types of firearms. From this time forward there was a dual development of more effective gunpowder and increasingly efficient types of weapon to use it. Many types of firearms such as the breech loader and the revolver would have become fully practical far earlier than they did had the gunpowder at the time been a better product, but it was not until the 1880s that any major change in the composition of a propellant charge for firearms was made. When this change—the development of nitrocellulose—came about, it revolutionized firearms design and function to the same degree that, five centuries before, gunpowder had revolutionized man's capabilities for waging war.

Production of gunpowder in England dates back with certainty to 1345, but not until the reign of Elizabeth I were powder mills, as we understand them today, established, the famous mill at Waltham Abbey dating from about 1561, and that at Faversham (extant today and on view to the public) from the same period. The first powder mill in the United States dates from 1675 and was located at Milton, Massachusetts.

Gunpowder is a physical combination rather than a chemical compound (as is modern 'gunpowder' or nitrocellulose) made by mixing three ingredients together in a moist, pasty state. The

Below left: this German drawing of 1450 depicts a firemaster, on the left, who is supervising the pounding of gunpowder. Each man pulls down a heavy wooden beam mounted on a spring into a trough, where it crushes the ingredients which will be made into gunpowder.
Below right: the firemaster, in another picture from the same work, is overseeing the weighing and mixing of the essential ingredients to make up the composition of gunpowder. The ingredients are then mixed together in a moist, pasty state. A rope fuse is being fitted to each completed 'grenade'.

resulting mass is then compressed under tremendous pressure into blocks, which are then broken up and reduced to various granular sizes. These are then put through sieves to sort the sizes according to their intended use. The ingredients are potassium nitrate (saltpetre), charcoal and sulphur. Great importance was attached to the quality of the ingredients, particular attention being paid to that of the charcoal, which was considered critical to the ultimate strength and explosive force of the powder. Charcoal from a variety of dogwood was used for military small arms and best quality sporting gunpowder, while willow and alder charcoal went into ordnance and cheaper grades of sporting powder. There was a great deal of variation in the quality of commercially produced gunpowder in the eighteenth and nineteenth centuries, depending largely upon the market for which it was intended.

Until about 1650 the proportions of the ingredients varied considerably. The charcoal and sulphur were present in much greater proportions than subsequently, each usually about one quarter of the total, and sometimes as much as one third of each ingredient. From about 1650 the proportions stabilized at the point where they were, with few variations, to remain: 75% saltpetre, 15% charcoal and 10% sulphur.

Gunpowder is a combustible, not an explosive. By varying the process of manufacture and the proportions of the ingredients it can be made to suit all forms of weapon or other commercial uses. For these reasons, despite its supersession by nitrocellulose for small arms and most other weapons, it remains in production for the many valuable industrial functions which more modern chemical compounds cannot economically fulfil.

Left: this mid 17th century drawing shows 'fire bombs' attached to arrows which are about to be shot from a bow and matchlock cannon. The fire bombs probably contained gunpowder as an incendiary agent, but they could equally have been impregnated with oil. The fuse attached to the bomb was lit before firing.

Below: Leonardo da Vinci's illustration depicts mortars discharging explosive shells and shrapnel.

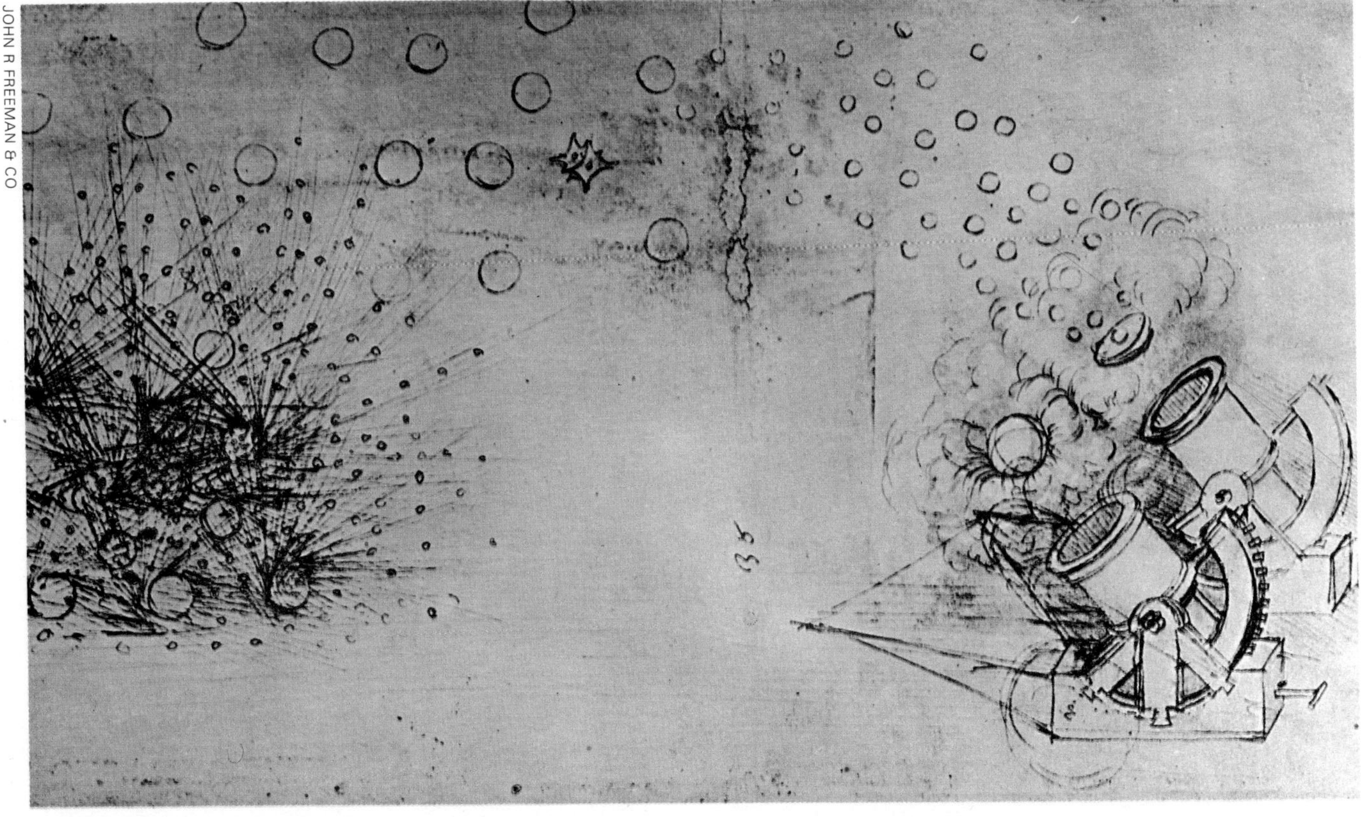

RADIO TIMES HULTON PICTURE LIBRARY

THE INDUSTRIAL REVOLUTION

The Industrial Revolution took place when metallurgy and machine-building were far enough advanced so that machines could work by themselves continuously, as well as providing mechanical advantage. Industrial productivity rose with the development of the steam engine, which changed economics, politics, family life—the whole fabric of human existence. When electricity came into use, the basis of invention became scientific rather than empirical, and so more complicated.

TO THE
TELEGRAPH OFFICE
ON DEPARTURE PLATFORM
Pears
Nº 221

The flying shuttle

Of all inventions, those connected with the textile industry have probably been the most controversial: they sometimes led to riots and the passing of laws prohibiting their use. Throughout the Middle Ages and later, spinning wool fibres into yarn and weaving it into cloth were essential industries which produced basic goods for trade, playing a great part in the economies of nations.

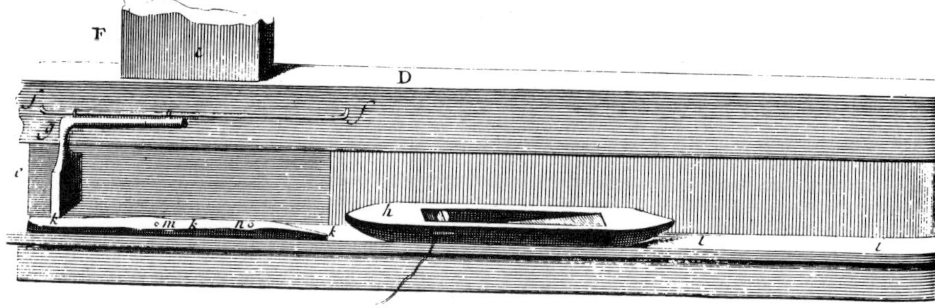

In most countries, wool was most important, on account of its cheapness, until the American plantations made cotton supplies more widely available. In Britain, the 'soft' water and unsuitability of moorland for anything other than sheep farming made the northern counties of Lancashire and Yorkshire important centres of the woollen cottage industry.

The process of weaving wool is basically a simple one, and the remains of looms have been found on neolithic sites. The general design of the horizontal loom was well established in Europe by the 13th century, but few changes, other than in quality of manufacture, were made until the 18th century. Various attempts were made to produce automatic looms, worked by the simple movement of a lever, and in the 17th century such devices were indeed made, though mostly for narrow strips or ribbons of material. Riots or mechanical problems resulted in these looms failing to become popular.

The invention which changed this picture was John Kay's flying shuttle. It was not a completely new loom, or an automatic loom, but simply a device which could be added to existing looms to improve their efficiency.

In a loom, a row of parallel threads, the *warp*, is arranged into a *shed*, with alternate threads lifted up to make a space through which a *shuttle* carrying another thread, the *weft*, passes. When the weft reaches one end of its travel, this thread is pushed against the previously woven ones by means of a comb or *reed*. The warp threads are then moved so that the other set are lifted, and the shuttle is passed through the shed again, this time in the opposite direction. So the weaving progresses.

Even early European horizontal looms had treadles arranged so that each shed was formed automatically by pressing a lever by foot. The shuttle, which carried the wool wound on a bobbin at its

The print below shows a pair of looms before the introduction of Kay's flying shuttle. The width is limited by the distance a man can stretch his arms. Above is shown Kay's shuttle device itself, with picker, g, and shuttle, h.

centre, was passed or more literally thrown from one side of the shed to the other. This limited the speed of operation, and also limited the width of material which could be woven.

John Kay came from Bury in Lancashire, where he started his own business making wire reeds to replace the split cane ones used until that time. Kay's reeds resulted in better cloth and fewer yarn breakages; he also made and patented several other devices to improve spinning and weaving, but it is for his flying shuttle that he is remembered.

Instead of being thrown through the shed by hand, in Kay's method the shuttle was flung across by an impulse given at either end of its travel. At each end was a *shuttle box*, containing a spindle along which a block called a *picker* could slide. Each picker could be pulled by a cord so that it could be suddenly jerked, giving the shuttle, which rested beneath the spindle, a violent push making it fly across to the other shuttle box.

The two shuttle boxes were linked by a batten over which the shuttle ran, sliding over the lower yarns. To make the shuttle fly more easily, John Kay equipped it with rollers; and to make the yarn wind off it more easily, he made the bobbin conical so that the wool was pulled off the top of the cone rather than being unrolled from a spindle. As the shuttle reached the end of its travel, it was slowed down by a groove, thus preventing it from rebounding so far that the picker would not hit it.

Kay's invention, patented in 1733, had an immediate effect on the industry. It meant that more weaving, of greater width, could be carried out by fewer people. The traditional looms had something like half the speed, and were limited to a width of 30 inches (76 cm).

Far from being welcomed as a way of increasing output, Kay's device was unpopular because of the fears of unemployment, and the legal battles which ensued lost him money rather than bringing him wealth. In 1747 he tried his luck in France but after initial success his efforts failed. On a visit to Bury in 1755 he was attacked by a mob who destroyed one of his looms, though by that time the idea had already caught hold. Kay died a poor man in France, his death unnoticed and unremarked.

Kay's flying shuttle is regarded as the first step towards the Industrial Revolution. When looms were mechanized some 50 years after his invention, the principle of the flying shuttle succeeded where other methods had failed.

C M DIXON

MICHAEL HOLFORD LIBRARY

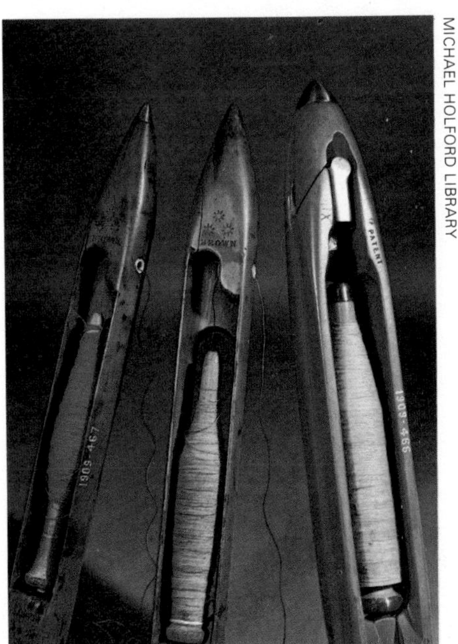

Left: although modern machine looms no longer have the flying shuttle in a recognizable form, this hand loom currently in use in Tunisia has a slightly adapted version of the system. The cord, to which the pickers are attached, is tugged either way by means of a pair of cords which pass over pulleys at the top of the loom.

Above: three Kay-type shuttles, taken from old looms. Earlier shuttles had rotating bobbins which could snag and break the yarn when used at speed. These fixed bobbins unwound much more rapidly. The first designs had wheels, but it was found that these were unnecessary.

The seed drill

For thousands of years the traditional method of seeding was to broadcast the seed by hand, and the land lay fallow every other year. Broadcasting was highly inefficient, as much seed was taken by birds or subsequently choked by weeds. This practice, however, was continued well into the 19th century although developments in seed drills had been made in the 17th century.

But crude seeding devices were known in China and Arabia long before this. The Chinese invention took the form of a wheelbarrow with a hopper for the seed which was delivered from three spouts into crude furrowing devices. The Sumerians actually placed a hopper on the plough beam behind the share, the seed dropping through a tube into the furrow immediately it was turned.

The Italians can probably claim to be the first European inventors for whom there is some documented evidence, a patent being filed by C Torello in 1566. The next seeding device, by T Cavalini of Bologna, was described in 1602 as being 'very useful for flat country'.

At the same time English inventors were working on seed drill designs. Alexander Hamilton patented a machine in 1623 but it was probably not practical. In 1669 Worlidge described a force feed seed drill which also had a hopper for spreading manure simultaneously. During this period a German called Locatelli invented a spoon feed seed drill which was attached to a Spanish plough. The seed box was divided in two, one part holding the seed and the other the seed dropper.

But the first really practical device was Jethro Tull's invention. It is claimed that the high cost of seed, which in 1701 rose from five shillings to twelve shillings, encouraged him to design an efficient seed sowing machine. Apart from his interest in agriculture, Tull was an accomplished musician, so it was not surprising that he hit on the idea of a seeding mechanism based on the grooved tongue and spring of an organ soundboard. His first seed box was made by a well known Soho instrument maker and tried out on the frame of a wheelbarrow.

The mechanism was made by enlarging part of the wheel shaft and making it toothed. As the wheel turned the teeth rotated against a spring-held tongue and seed was released from the seed box to the seed dropper at regular intervals: the number of teeth around the shaft governed the spacing of the sown seed.

Tull's first seed drill, used for sowing sainfoin and wheat, made the furrows, drilled the seed and finally covered it. Tull made several variations of his drill

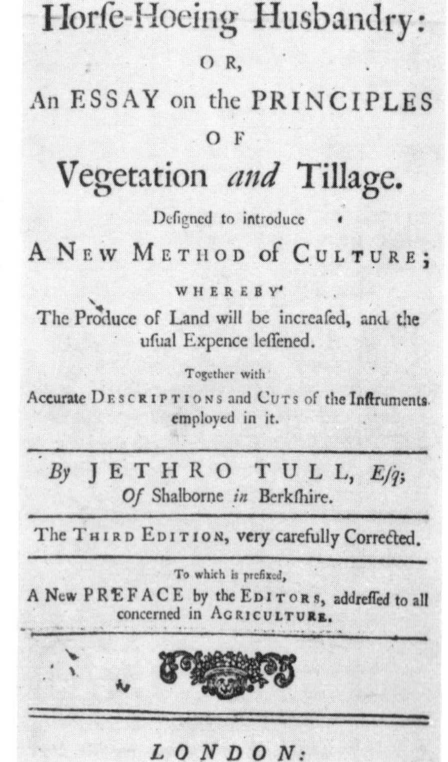

Horſe-Hoeing Husbandry:
OR,
An ESSAY on the PRINCIPLES
OF
Vegetation *and* Tillage.
Deſigned to introduce
A NEW METHOD of CULTURE;
WHEREBY
The Produce of Land will be increaſed, and the uſual Expence leſſened.
Together with
Accurate DESCRIPTIONS and CUTS of the Inſtruments employed in it.

By JETHRO TULL, *Eſq*;
Of Shalborne *in* Berkſhire.

The THIRD EDITION, very carefully Corrected.

To which is prefixed,
A New PREFACE by the EDITORS, addreſſed to all concerned in AGRICULTURE.

LONDON:

MUSEUM OF ENGLISH RURAL LIFE

Above right: the title page of an early edition of Jethro Tull's book on horse hoeing. The book, first published in 1733, contained many original ideas but did not greatly influence 18th century farming.

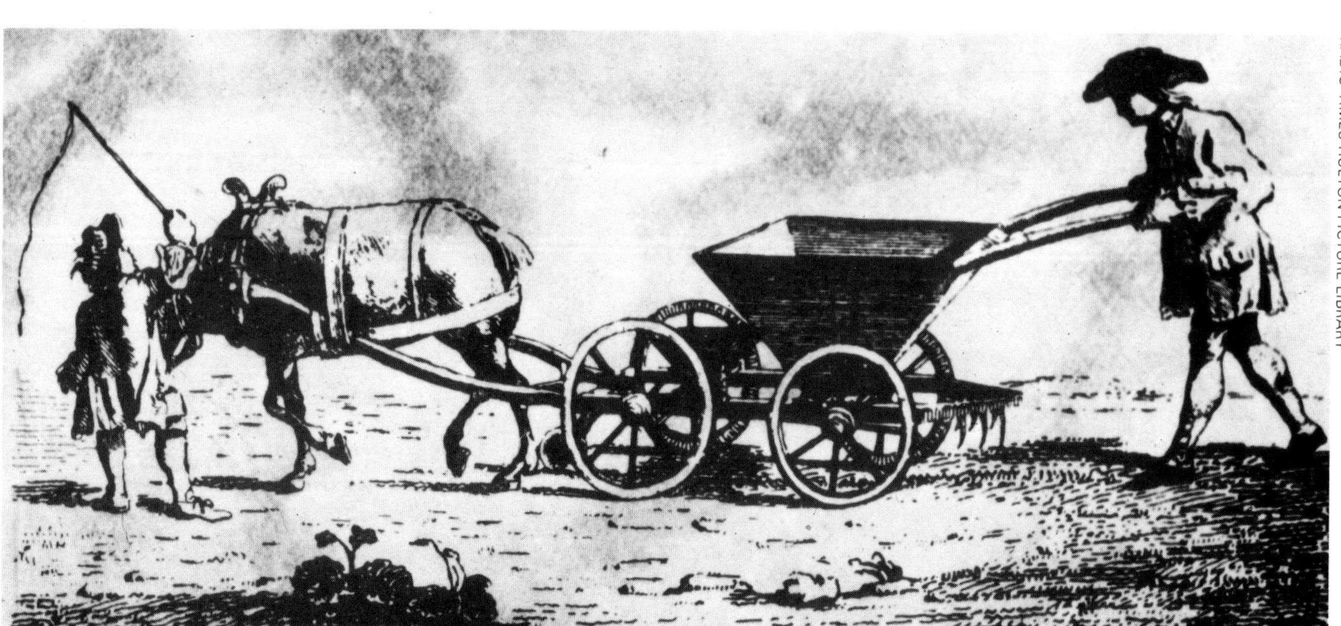

RADIO TIMES HULTON PICTURE LIBRARY

Below: a drill-plough invented in 1745 by William Ellis. It was equipped with two hoppers, one for seed and the other for manure, but according to contemporary reports it was temperamental and did not always work properly.

but all were based on the same type of seed dropper.

His most advanced design, described in his book *Horse Hoeing Husbandry* published in 1733, sowed three rows of seed and was drawn by one horse. The three coulters were narrow and shaped to enter the soil readily. At the rear they had narrow passages, open behind, for guiding seed from the funnels above to the furrows in the ground. These coulters and the framework supporting them rested on the ground and not on the four wheels of the machine. The drill had four wheels: two large wheels in front carried a seed box and dropper unit which fed the centre coulter, while two small wheels at the rear carried droppers and seed boxes supplying two other coulters, which were placed 14 inches (35.6 cm) apart and some distance behind the centre coulter so as not to interfere with it.

Tull also went into the economics of crop growing. He found that it cost £4 10s to produce 30 bushels of wheat per acre: this figure included the cost of three bushels of seed, tillage, hand weeding, rent of land during the fallow period, manure and reaping. With a drill, however, it was possible to use less than a third the amount of seed needed previously and yet increase the yield, recovering 45 to 80 bushels per acre at a cost of only 15 shillings. This amount included the seed, tillage, drilling, weeding, uncovering, additions of brine and lime, and reaping.

Although a writer of 1770 commented 'the spirit of drilling died with Mr Tull', other inventors continued to develop their own designs for sowing grain, beans or potatoes. Only the most progressive farmers, however, used drills.

Of interest was Cook's spoon and cup feed drill patented in 1782, and later improved by Baldwin and Wells. They introduced a sliding axle, thus enabling the distance between the wheels to be extended, allowing more cups and coulters to be fitted. The coulters were attached to separate self-regulating levers which enabled them to accommodate any differences in levels in rough terrain.

By the 1860s seed drills were of modern design: each coulter was fitted with an independent lever to adjust it to different widths; gear drive was incorporated as opposed to the earlier belt drive; they had swing steerage; and an improved manure box for applying fertilizer simultaneously with the grain.

Right: a seed drill of the early 19th century. Notable features are the gear drive from the main wheels and the provision of a steering mechanism.

Below: a drawing of Jethro Tull's seed drill and its component parts. The machine had three coulters for sowing the seed, a central one operated by the large main wheels and two others operated by the smaller rear wheels.

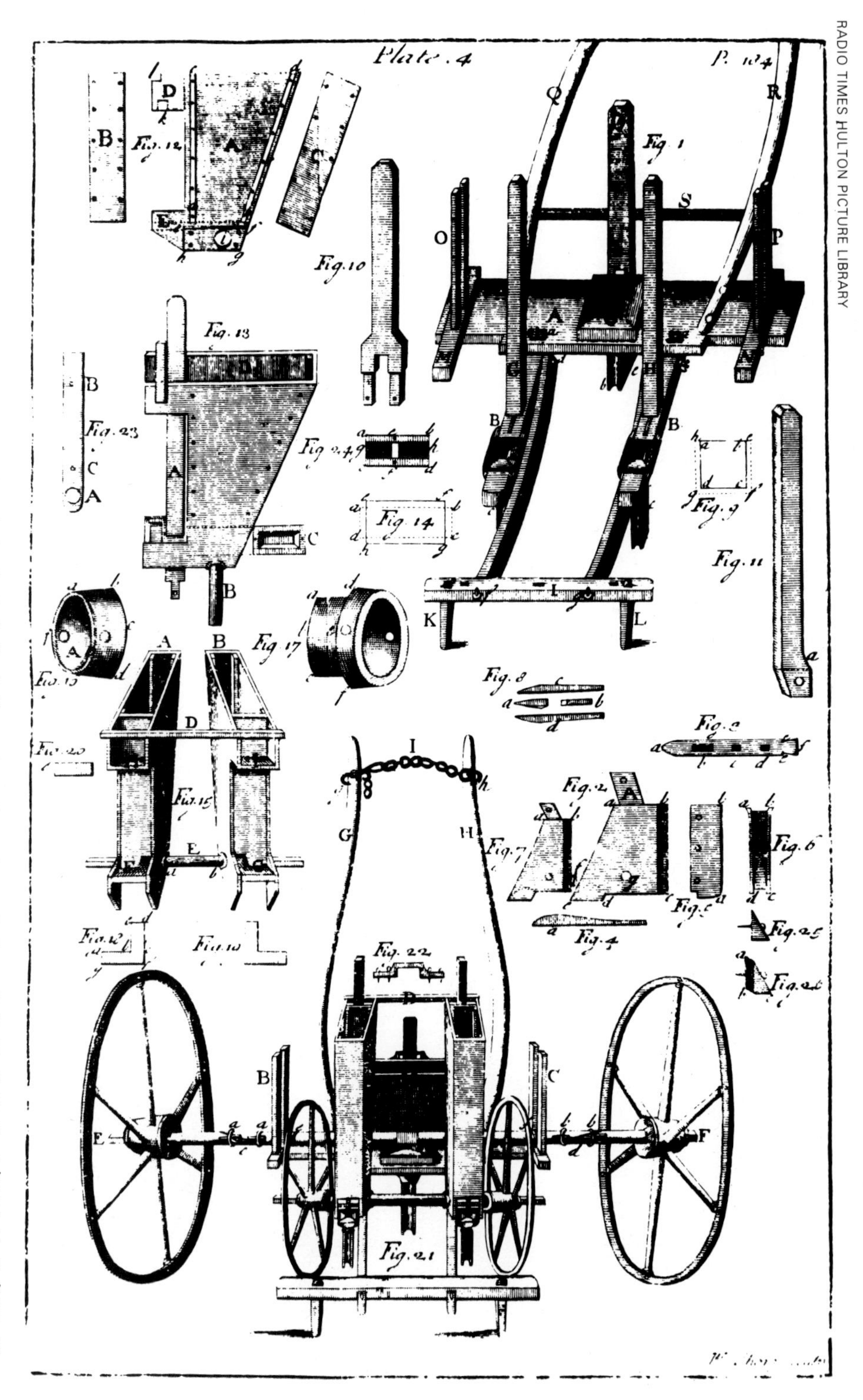

The chronometer

Even as late as the 18th century, navigation at sea was a somewhat uncertain business. With the aid of a quadrant, an angle measuring device, navigators could work out their latitude reasonably well by measuring the angle above the horizon (*altitude*) of the Pole Star—always within a degree of the angle of latitude—or, more accurately, by measuring the altitude of the Sun at its highest daily point above the horizon. This was the moment of local noon, and the latitude was worked out from tables of the Sun's motion.

A common method of navigation was to sail to the latitude of the destination required, then sail along the line of latitude until land was reached. Navigation was largely by 'dead reckoning'—keeping track, as much as possible, of distance travelled in each compass direction. If both distance and direction could be measured with high accuracy, this would be a perfectly good method, but in practice such factors as ocean currents and magnetic variation made the method hazardous, and sometimes fatal.

What was needed was a method of finding longitude at sea. Although latitude is quite easy to find the fact that the Earth rotates means that there are no fixed points on which to base observations. One needs to know the time very accurately in order to allow for the Earth's rotation: what we refer to as 'Greenwich Mean Time', the world's time standard, is no more, in principle, than the sundial time at Greenwich—that is, the position of the Earth's rotation relative to the Sun. But when the Sun is at its highest at Greenwich, it is still five hours from its highest as seen from the coast of America.

Although 18th century clocks were capable of giving satisfactory results on land, the problems of using them at sea were very great. In particular the motion of the ship produced sudden accelerations which affected the pendulum, and the extreme temperature variations were likely to alter the rate of the time-keeping mechanism.

Following a disaster off the Scilly Isles, near the tip of Cornwall, in which 2000 lives were lost, the British government in 1714 offered a reward of up to £20,000 to the first person to provide a useful and practicable method of finding longitude at sea. This was to be awarded by a Board of Longitude if the method should be accurate to within $\frac{1}{2}$° of longitude, which is 30 nautical miles; lesser amounts for less accurate methods were also offered, and up to £2000 for experimentation was available.

One method available was to observe the Moon's position relative to the stars, comparing this with tables. The Moon moves round the Earth in 29 days, so it acts as a timekeeper which is independent of the Earth's rotation. The drawbacks of this method were that angles had to be measured to within a minute of arc, about $\frac{1}{30}$ of the Moon's diameter as seen in the sky, and that accurate lunar tables had to be compiled. Such tables were eventually completed in 1755 by a German, Tobias Mayer, though the difficulty of making the observations still limited accuracy.

Below: Harrison's first timekeeper, known as H1. The two rods, each with a ball at either end, oscillate to and fro. The gridiron rods can be seen lying horizontally in the centre of the mechanism: they are linked to the spring tensioning. The clock is in full working order, and can be seen operating at Greenwich, England. At right is the improved version with remontoire, H2, made in 1739.

NATIONAL MARITIME MUSEUM

The most famous applicant for the reward was John Harrison (1693–1776) a self-taught Yorkshire clockmaker. He was very concerned with the problems of clock accuracy, and one of his early inventions was the *gridiron* pendulum. This uses rods of brass and steel, which expand to different lengths for any rise in temperature. The gridiron pendulum has a row of rods connected in such a way that the expansion of one is compensated for by the other so that the pendulum is always the same length.

Harrison's first attempt at a sea going timekeeper was finished in 1735 and is generally known as H1. To overcome the ships' motions, he used two large balances, arms weighted at each end, acting against springs in such a way that any external motions would have opposed effects, and would be cancelled out. He also used a gridiron mechanism constructed in such a way that the effects of temperature would be magnified, not reduced, to adjust the tensioning of the balance springs.

This first machine was large and open to the air but was found to be accurate when given a short sea trial. Harrison then began work on an improved version, H2, on the basis of what he had learned: the result, though better, was still not enough to qualify him for the full reward.

Harrison's timepieces were great triumphs of his art, involving many new ideas such as a 'grasshopper' *escapement* (which links the time-keeping mechanism to the power source, in this case a wound spring). This was designed to have as little friction as possible in its operation. Many of Harrison's innovations are concerned with reducing friction: for some bearings, *lignum vitae*, a naturally oily wood, is used.

Harrison went on to make another clock, H3, which replaced the gridiron system with what he called a 'thermometer kirb'. We know this today as the bimetallic strip, used in thermostats. It consists of a strip of brass and a strip of steel joined together so that as the temperature changes, the different expansions of the two metals cause the combined strip to bend.

This movement is used to regulate the clock so that it operates at a constant rate whatever the temperature.

Still unsatisfied with the accuracy of the clock Harrison, now joined by his son William, abandoned many of his earlier design ideas, though keeping the bimetallic strip. The result of this, H4, was much smaller yet even more accurate than its predecessors: it is about the size of the average alarm clock. Instead of the large balances he used two small wheels, similar to those to be found in most ordinary watches and clocks today, controlled by a single balance spring whose effective length is adjusted by the bimetallic strip. A *remontoire* is used to provide constant force: this is rewound by the main spring every 7½ seconds, so that variations in the main spring tension do not affect the power available.

This chronometer was taken on a long sea trial to the West Indies, as specified in the requirements of the Board of Longitude. In five months the total error amounted to less than 15 seconds. This was well within the limits required by the Board but, being careful with the government's money, they paid only half the award. The other half was not paid until Harrison had given full details of the clock's construction, and a replica had been made by another clockmaker.

Even so, the chronometer did not come into widespread use until the 19th century, since its high cost offset the greater accuracy possible.

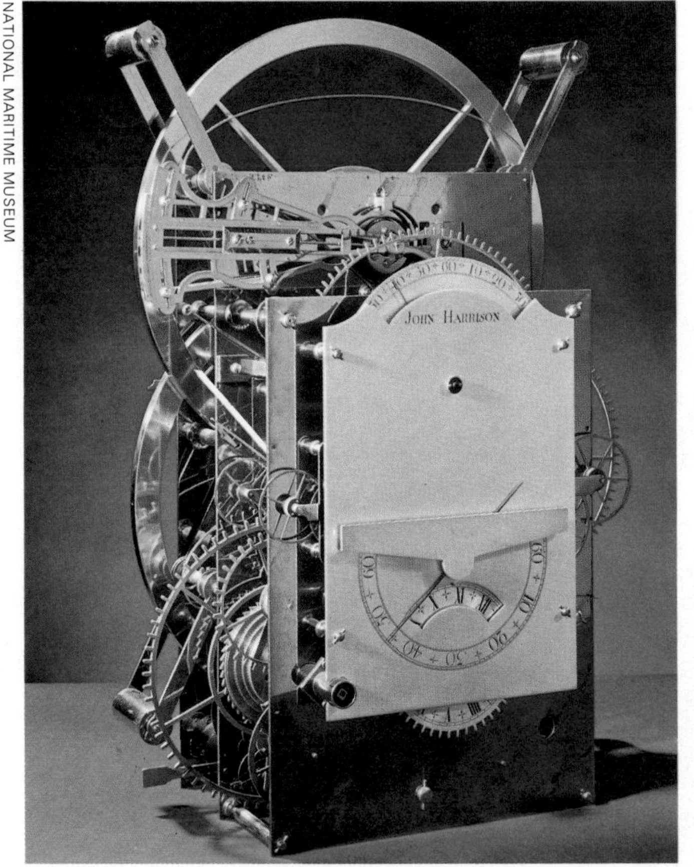

NATIONAL MARITIME MUSEUM

Left: after a long interval, Harrison produced this timekeeper, now called H3, in 1757. It has two large balance wheels instead of arms, and uses the 'thermometer kirb' or compensation curb system of temperature correction. A bimetallic strip bends as the temperature changes, so moving curb pins along the balance spring.

Below: the famous H4 (left) of 1759 and, on the right, Kendall's 1769 copy. Only in 1773, when Harrison was 80, did he receive the full £20,000. H4's accuracy of 15 seconds in five months compares with one second per month achieved by modern quartz crystal watches. The timekeeper is five inches (13 cm) in diameter.

Electrostatic generation

Modern electrical research started in 1600 when William Gilbert, court physician to Elizabeth I and amateur scientist, tried to distinguish between the attractive properties of a magnet and those of a rubbed insulator such as amber, glass and sulphur. He simply rubbed these by hand. Some sixty years later, the first electrostatic generator was devised by Otto von Guericke, the scientifically-minded mayor of the small German town of Magdeburg. His machine consisted of a sulphur ball 'about the size of a child's hand' both rotated and rubbed by hand, and with it he observed many of the properties of electrostatics that are known today. However, Guericke did not see these investigations in terms of electrical experiments, or the machine as an electrostatic generator, but only used it to demonstrate his cosmological theories of the universe. He observed that in nature there existed a number of forces, including attraction and repulsion, and he could create these by rubbing his sulphur globe. He equated the attraction produced in this way with the Earth's gravitation. It was only during the eighteenth century that the electrical nature of these experiments was recognized and it was assumed that they had the same significance for Guericke.

It is often stated that in 1675 Sir Isaac Newton improved the electrical machine by replacing the sulphur globe with one of glass, but this is based on a historical confusion. There exists no evidence whatsoever that he employed such a device in any of his experiments, although his interest in electrical phenomena is historically well-known. He demonstrated that glass was a good 'electric' (meaning a good insulator or *dielectric*) and had strong attractive properties when rubbed.

The first real frictional electrical machine was made in 1706 by the elder Francis Hauksbee, a maker of scientific instruments and the unofficial Curator of Experiments at the Royal Society of London. It was the outcome of his solution of a very puzzling observation made by the French astronomer Jean Picard some thirty years before. Picard noticed that when he moved his barometers in his observatory, a glow sometimes appeared above the level of the oscillating column of mercury. In a careful series of experiments started in 1703, Hauksbee demonstrated that the glow was caused by the friction of the mercury on the glass barometer tube, which produced electricity, and that the same phenomenon could be obtained by rubbing a rotating glass globe with a woollen cloth or the open hand. His device was produced commercially both in England and in Europe, but it did not become immediately popular, and appears to have been used primarily for replicating Hauksbee's experiments and not for original research.

During the 1730s, major advances in electricity were made by experimenting with rubbed glass tubes and rods; a method also used by Hauksbee. The frictional or *triboelectric* generator only became popular in the second half of the 18th century, mainly as the result of public experiments made with this device by a number of German enthusiasts, which also led to its improvement. The glass was rubbed by a leather or silk

INSTITUTE OF ELECTRICAL ENGINEERS

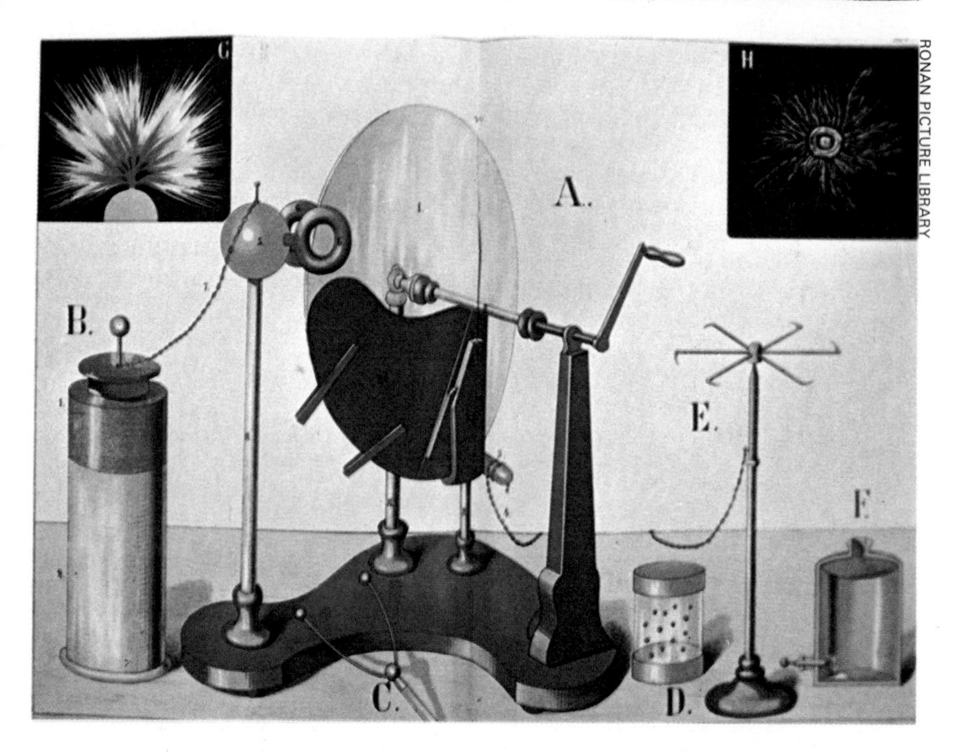

RONAN PICTURE LIBRARY

Top: Otto von Guericke's sulphur ball machine of 1660. Fig VI shows his famous feather experiment. The feather at 'a' remains in this position after it has been attracted and repelled by the charged ball. Above: electrostatic machine with glass plate (A) charging a Leyden jar (B).

SCIENCE MUSEUM

cushion whose pressure could be adjusted, and the charge was collected by an insulated metal tube, called the 'prime conductor', which was supported close to the surface of the rotating glass. There were many different designs, but all the machines worked the same way. Rubbing the glass gave it a positive charge, which was neutralized by the pointed ends of the prime conductor terminating near the glass surface. This conductor was insulated, so it too became positively charged.

A typical portable electrical machine was designed by the London instrument-maker John Reid in 1765, specifically for medical treatment. The chemist Joseph Priestley, best known for his independent discovery of oxygen, also designed a generator which was copied by Benjamin Franklin in America. The largest machine ever constructed was made in Holland by the Anglo-Dutch instrument-maker John Cuthbertson in 1783. Its huge twin glass plates produced discharges two feet (61 cms) long, that is, about 500,000 volts but of very small current. When Volta discovered the 'voltaic pile' in 1799 to 1800, interest switched from electrostatics to current electricity and the frictional generator became obsolete. Renewed interest in high energy electricity during the second half of the 19th century led to the design of a new generation of electrostatic machines known as influence or induction machines of which the Wimshurst, designed by the British engineer James Wimshurst in 1881, was the most popular. Although the frictional machine could still be purchased, it was only used for demonstrations in school and university laboratories. In the 18th century, however, this device had been important in the progress of electrical discoveries.

Left: J Priestley's globe electronic machine (about 1769). The globe is rubbed by the cushion at its base. The prime conductor arrangement is not shown here.
Below: Hauksbee's machine rubbed by hand.

JOHN R FREEMAN & CO

Volta's pile

SNARK INTERNATIONAL

MARY EVANS PICTURE LIBRARY

Volta demonstrated his new battery to Napoleon in 1801 (above left), whereupon Napoleon ordered a giant 600 element pile to be built. This was installed at the Ecole Polytechnique (above). The modern French word for battery is still 'pile'.

There were two ways of obtaining electricity by the end of the 18th century: directly from an electrostatic machine, generating static charges by means of friction, or from the storage of such electricity in a Leyden jar, the bulky forerunner of what today would be called a capacitor. There was no source of current, as distinct from static electricity, and indeed the existence of current was not suspected.

The invention which provided a source of current electricity was triggered off by a discovery by Professor Luigi Galvani of Bologna University, Italy. He found that frog's legs would twitch when two dissimilar metals were in contact with them. It had been known for some time that electricity caused contractions in muscular tissue, and Galvani thought that the animals' bodies themselves stored electricity, the metals simply serving to conduct it.

Not everyone was convinced by his theory of 'animal electricity'. In particular, Alessandro Volta, professor of physics at Pavia University, disagreed strongly with Galvani. Volta had already carried out much research in electricity and had invented several instruments whose purpose was to accumulate charge or to measure electrical effects exactly. A vigorous controversy broke out between them, complicated by political differences, for both were involved in the struggles of those revolutionary years. In Volta's opinion, the contact of two moistened and dissimilar metals produced these strange results: such metals could, he wrote, 'by themselves, of their proper virtue excite and dislodge the electric fluid from its state of rest'. So he began a series of experiments on himself, reviving an idea which had been tested some years before by a philosopher who wished to explain pleasant and unpleasant sensations. He took a piece of tinfoil and a silver coin so as to pinch his tongue between them, while they were in contact at the edges. He felt a distinct sour taste: when the pair similarly touched his eye, he saw flashes. Gradually he abandoned the idea of even a passive animal electricity stimulated by the metals. Any moist substance would do. There were 'dry conductors'—such as metals and charcoal—and 'wet conductors'. Any contact between the two classes would 'agitate or disturb the electric fluid'. And if one of either class were placed between two of the other—or two within one—a continual circulation would be set up, which would stop only when the circuit of the three was broken.

What if these sets of three should be added, one to another? He took a silver coin, then a zinc plate cut to fit over it; then a piece of wet pasteboard; then another silver-zinc-wet pasteboard combination; then another. The more couples, the stronger the effects—soon he was constructing towers of twenty,

MICHAEL HOLFORD LIBRARY

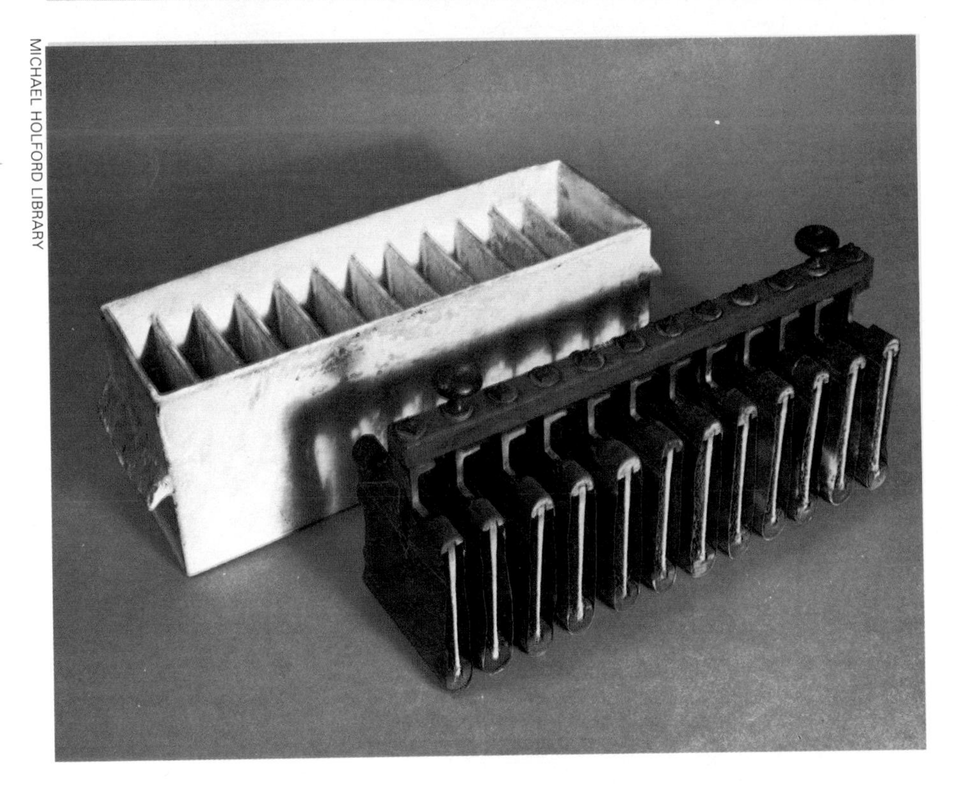

Top: this shows the design of Volta's crown of cups, the terminals being the ends of the bows at the front.

Above: subsequent batteries, such as Wollaston's shown here, used plates—a similar design to the modern car battery.

thirty, sixty sets, kept in position by rods. If he touched the two ends of the pile, he felt a shock, and the higher the pile the fiercer the shock. Sparks appeared when wire leads were attached at each end. And the best of it was that this flow was continuous: it did not discharge itself immediately and then have to be recharged, like all earlier electrical devices. He tried touching nose, eyelids, ears, forehead with the wires; at each organ there was an appropriate sensation, and the continuity of these sensations proved to him 'an endless circulation, or perpetual motion' of his current. Sometimes he found that in the larger piles, the water was squeezed out of the lower pasteboard layers, so he designed what he called a 'crown of cups': they were filled with a saline solution, which he had found was better than pure water, and linked to each other by means of 'metallic arms, or bows, one arm of which is silver . . . and the other of zinc. The ends of these bows are plunged into the liquid always in the same successive order, the silver ends being all on one side and those of zinc on the other'. A bow could be made by wrapping one end of a strip of zinc around the handle of a teaspoon.

Volta announced his discovery in a letter to the President of the Royal Society in March 1800. Within a month a voltaic pile in London had been used to decompose water (*electrolysis*); solutions of metallic salts could be similarly decomposed. Over the next few years many such electrolytic researches were undertaken, culminating in the isolation of sodium and potassium by Humphry Davy in 1807. More satisfactory arrangements of the pile were introduced, with plates laid sideways in boxes; cheaper metals could be substituted, provided they occupied appropriate places in an electromotive series that was now revealed. Volta himself went to Paris and was feted by Napoleon, who ordered an enormous battery pile of 600 plates, while in London the Royal Institution soon collected subscriptions for another with twice as many. Volta's pile had given electricians for the first time a steady current. Despite some shortcomings in his interpretation, he had invented the first electric battery, the essential precondition for all research on electrodynamics ever since—and for the use of electric energy.

Machine tools

In the 18th century many of the mechanical components of pioneer industrial technology, such as water wheels, textile machines, mining and metallurgical equipment, early railways and their wagons and so on, were made wholly, or very largely, of wood. Increasingly, however, in order to meet the need for higher strength and greater durability of moving parts, especially where stresses were high and operating speeds faster, metal parts had to be introduced. The change over to accurately constructed, more reliable and more complex machines made of iron and (for special components) steel was one of the most significant features of the Industrial Revolution, and it demanded the development of a new breed of machines which now occupy a position of central importance in modern technology—machine tools.

Machine tools have their origins in the specialized needs of the makers of furniture, clocks, instruments and guns. For centuries furniture makers and others working with wood had used simple lathes to turn various components of elaborate shape. The gunmaker applied powered boring tools to the construction of cannon. Clock and instrument makers used a variety of gear cutting tools, grinding machines and lathes to precisely fabricate and graduate optical and mechanical parts. Notably there was a need to make screws of accurate and fine pitch (distance between threads), frequently of very small size. The screw cutting lathe was not a new idea in the 18th century any more than its existence was novel, but the new order of precision attempted, and achieved, by such instrument makers as Jesse Ramsden (1735–1800) was unprecedented.

Working on a scale far removed from that of clocks and instruments, boring engines derived from the machines used to fabricate cannon were introduced to *turn* (cut in a circular path) the iron cylinders of the steam engines which were rapidly adopted in the last quarter of the 18th century.

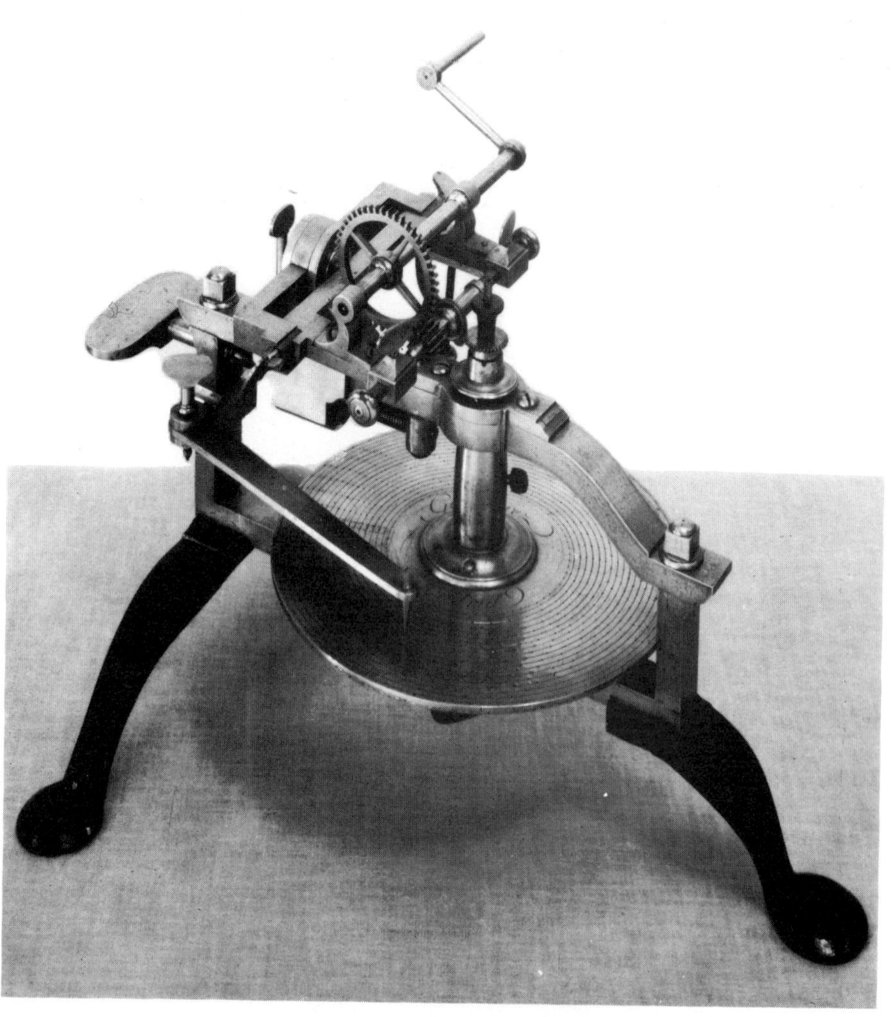

SCIENCE MUSEUM

SCIENCE MUSEUM

Above: a clock wheel cutter made in Madrid in 1789. The wheel being cut is near the top of the vertical spindle.

Right: an 18th century screw cutting lathe, used by clockmakers.

The all metal precision lathe capable of cutting fine screws and turning and facing the softer metals, the cylinder boring engine and other early machine tools were essentially one-off creations used exclusively in the workshops of their creators. The transition to production machine tools for sale to any factory or manufacturer who needed (and could afford) them began in England around the year 1800.

The prototype industrial lathe was evolved by Henry Maudslay (1771–1831). By synthesizing the best features of earlier practice, adding innovations of his own and insisting on the utmost precision, Maudslay radically improved the screw cutting lathe and extended its use to the working of iron. Engineers who had been trained by Maudslay, men such as Richard Roberts and Joseph Whitworth, introduced larger and more powerful lathes able to perform a wider range of operations. Theirs was the first generation of production industrial lathes embodying such features as variable speed drive, back gears, lead screws and traversing slide rests, as in a modern machine.

The lathe can be used for turning, screw cutting and facing, that is machining any flat surface which can be rotated in the lathe at the same time as the face is traversed by the cutting tool. The manufacture of many industrial and machine parts, however, required different techniques, particularly in the case of complex arrangements of flat surfaces with slots and grooves and any large plane surface which was too big to prepare on a lathe. The early machine tool repertoire was extended therefore by the invention of the milling machine and the planing machine.

The milling machine, an American innovation due to Eli Whitney in about 1820, made use of a rotary cutter fixed in position over a movable horizontal workpiece. The planing machine, using a fixed cutter applied to a surface moving to and fro, was developed in England, an early version being due to Richard Roberts in 1817. Soon after this date Roberts, among others, brought on to the market a gear cutting machine, a device which owed something to the wheel cutting machines of the clock maker. Once again, however, the transition was marked by a radical change of scale. The machine tool needed to cut industrial gears had to be capable of great accuracy and repeatable precision on large pieces of iron and steel if it was to meet the requirements of the more complex and higher speed geared drives which were becoming commonplace in every branch of industry and manufacture.

Machine tools revolutionized mechanical engineering. And more important still, their production became an industry in its own right. Ideas and concepts which first became a reality in Britain were quickly taken up elsewhere, especially in the United States, where the machine tool industry quickly established a pre-eminence destined to exert a powerful influence on future developments.

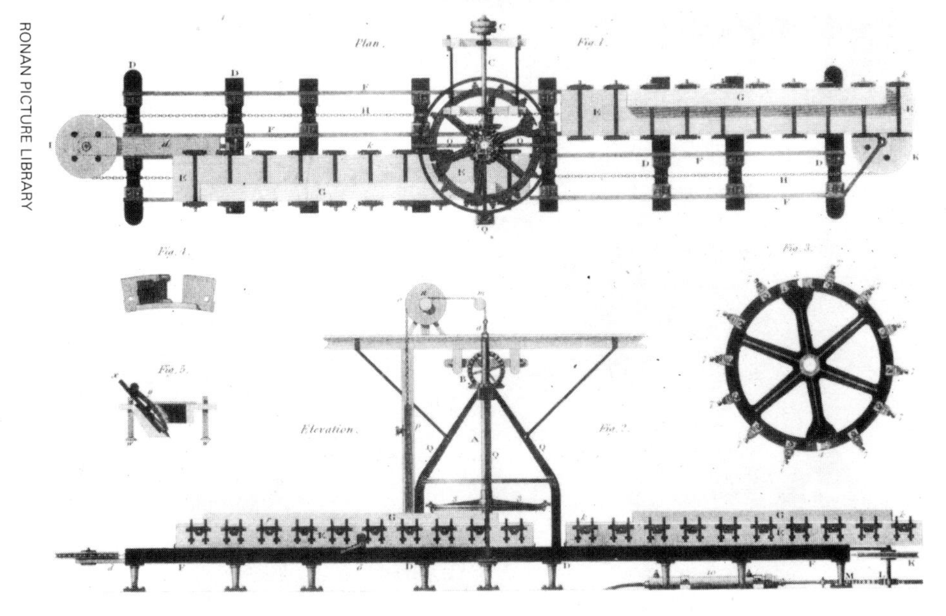

Top left: Eli Whitney's milling machine of the 1820s.

Left: a plan view and side elevation of Joseph Bramah's planing machine (1802).

Newcomen's steam engine

Newcomen's engine is now generally accepted as the first practical steam engine, and it was working over half a century before Watt's first engine. Thomas Newcomen was an ironmonger from Dartmouth, Devon, and he worked with a partner called John Calley. How much they knew about previous experiments with steam we can only guess; for example the French physicist Denis PAPIN built a model in 1690 to show how steam could be used to move a piston inside a cylinder. Some years later in Britain Captain Thomas Savery built a steam powered water pump which was partially successful, but it was not a steam engine in the usual sense; it did not have a reciprocating piston and so could not have been used to provide direct power for machines or vehicles. Savery, however, patented his steam pump in 1698 and his wide-ranging patent covered Newcomen's engine. As a result Newcomen and Calley entered into partnership with Savery in 1705.

After some experiments, Newcomen's first engine was built in 1712 on a site 'near Dudley Castle' (probably at Tipton in Staffordshire), according to a drawing made by Thomas Barney in 1719. Newcomen's engine made use of the fact that when steam cools down and

Below: Newcomen's atmospheric steam engine, showing pump for extracting water from a mine (right). From Architecture Hydraulique, published in Paris in 1739.

RONAN PICTURE LIBRARY

condenses back into water, its volume is greatly reduced. If this operation is carried out in a sealed container, a vacuum is created inside the container. Newcomen's container was a large vertical cylinder open at the top and fitted with a circular piston which could slide up and down. With the piston at the top of the cylinder, the volume below was filled with steam. A spray of cold water was then released into the cylinder in order to condense the steam and create a vacuum. Now the atmospheric pressure of the air acting on the upper face of the piston forced it downwards on a working stroke. The piston then had to be returned to the top of the cylinder ready for the next working stroke. Steam engines of this type were called after Newcomen (although he did not build them all): they were also called 'fire' engines and 'atmospheric' engines.

Newcomen's first engine was huge and occupied an engine house some 30 feet high. Dominating the layout was a 25-foot (7.6 m) long oak beam which was pivoted at its centre and rocked like a giant see-saw. Half of the beam was situated inside the engine house and this was connected by a chain to the piston inside its cylinder. The other half of the beam extended outside the building and was connected to rods which extended 150 feet down into a mine to power a water pump. The whole purpose of Newcomen's engine was to pump water out of flooded mines. At ground level, coal was stoked on to a fire situated beneath a domed boiler. The water in the boiler turned to steam, and this was transferred through a valve and into the large brass cylinder.

The pump side of the beam was deliberately made heavier to ensure that it rested in the down position and consequently the piston started at the top of the cylinder ready for the working stroke. The steam in the cylinder was condensed and the atmospheric pressure forced the piston down. This in turn lifted the pump rods on the other end of the beam and at the end of the working stroke their extra weight lifted the piston back to the starting position.

Newcomen engines were slow, low powered and very inefficient; nevertheless, they were better than anything else for pumping water out of mines and their use was wide-spread. Newcomen's first engine developed only 5 or 6 horsepower (3.7–4.5 kW), but later engineers improved on this performance. For example, John Smeaton, who built the Eddystone lighthouse, constructed several Newcomen engines including one with a cylinder 6 feet in diameter which developed about 75 horsepower.

Having discovered that the Newcomen engine was a reliable source of power, the engineers of the mid 19th century tried to adapt it to drive machinery which rotated—so far it had only been used for reciprocating pumps. One of the first attempts was made by John Oxley in 1763, who fitted a ratchet device to turn a shaft and raise coal from a mine. In 1780 a Newcomen engine was fitted with a crank by one of James Watt's rivals and this patent forced Watt into using a 'sun-and-planet' arrangement of gears instead. But Watt's engine with its separate condenser was very much more efficient than Newcomen's and rapidly replaced it—except in certain areas where coal was cheap and Newcomen engines remained in service even into this century.

Below left: one of Newcomen's atmospheric steam engines for draining mines, built in 1772 by John Smeaton.

Below right: the original engine near Dudley Castle. Note the fireman and the open fire door. Only the upper of the two pumps is shown.

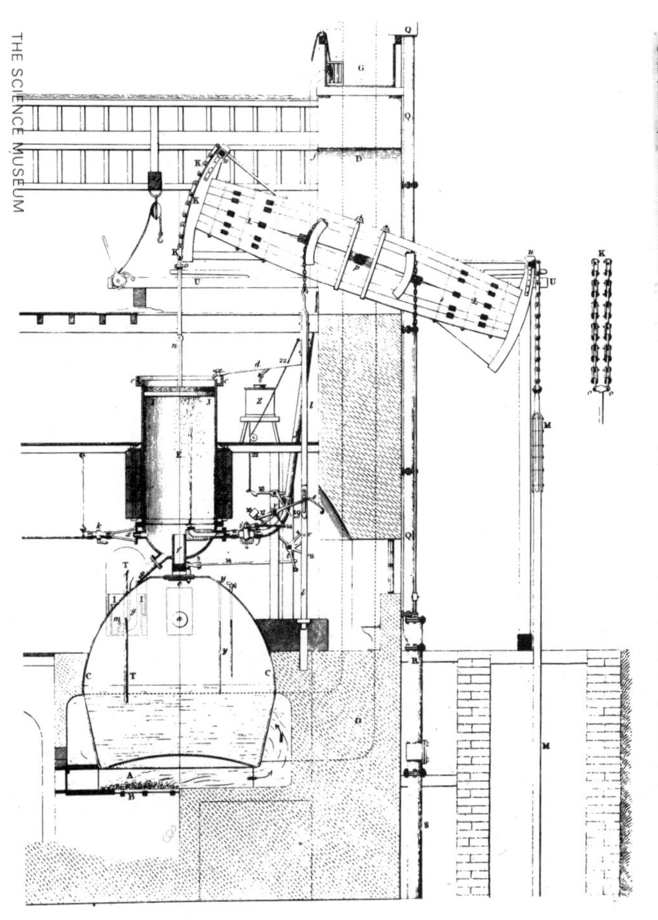

THE SCIENCE MUSEUM

Watt's steam engine

Contrary to the popular story, James Watt did not invent the steam engine after watching steam lift the lid of a kettle. In fact his inspiration for an improved steam engine came to him while he was walking across Glasgow Green on a fine Sunday afternoon in May 1765. Watt had been repairing a model Newcomen-type steam engine for the University of Glasgow.

Newcomen engines had been used for some 50 years to pump water out of mines. Steam was condensed in their cylinders, thus creating a partial vacuum beneath the piston; the atmospheric air pressure acting on the upper surface of the piston then forced it downwards on a working stroke. Watt realized that a large amount of heat was being wasted by successively heating the cylinder, to fill it with steam, and then cooling it to condense the steam. His solution to this problem was to fit a separate *condenser* which could be connected to the cylinder through a valve. The condenser could be kept cool while the cylinder remained hot. This was the idea which came to Watt on Glasgow Green and he demonstrated that it would work with a model. He patented the separate condenser in 1769, but many years were to pass before a practical engine was ready for sale.

Watt's first full size engine was built in Scotland at Kinneil House near Bo'ness but it was not a success. In spite of this, a Birmingham businessman called Matthew Boulton acquired a share of Watt's patent in 1773, and a very successful partnership began. Watt moved to Birmingham with his experimental engine and gradually overcame the practical problems. During 1776 the first engines were delivered to customers.

These Boulton and Watt engines resembled Newcomen engines in general appearance, and they too were used to drive pumps. The piston in their vertical cylinder was connected by a rod and chain to a seesaw beam, so that the working stroke pulled one end of the beam downwards. This movement naturally raised the other end of the beam, which was connected by rods to the pump. As the pump rods were heavier than the piston, their weight returned the piston to the top of the cylinder at the end of the working stroke. Watt's design incorporated two major improvements, however, and one of these was the

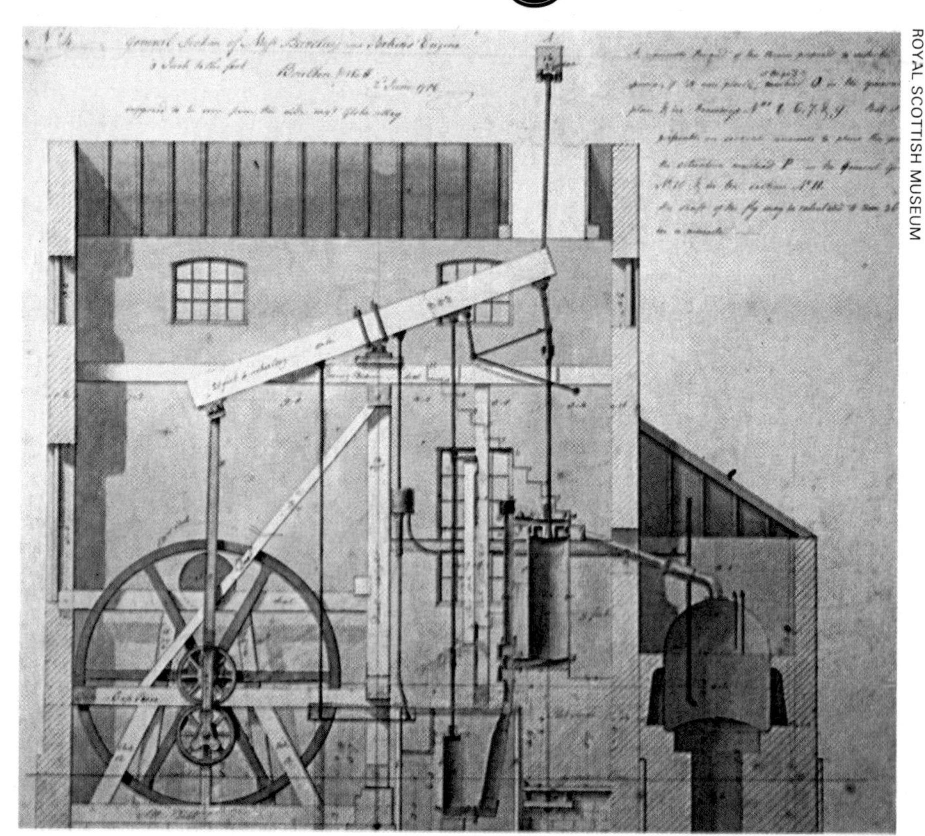

ROYAL SCOTTISH MUSEUM

Above: a Watt double acting engine of 1786, with sun and planet gears, separate condenser, and the pantograph linkage at the top of the piston connecting rod.

Below: this Watt engine was erected at Smethwick near Birmingham in 1777 and ran for 120 years until its removal, shown here, in 1898.

MANSELL COLLECTION

separate condenser. But he also covered over the top of the cylinder and admitted low pressure steam above the piston. This use of steam pressure instead of just air increased the downward force on the piston during its working stroke. As with most new designs, there were teething troubles, but the improvement in performance was quite remarkable. For example, one of the first Boulton and Watt engines to be installed for a customer required only one of the two boilers needed by the engine it replaced.

Watt continued to improve on his design, and almost every one of his early engines included a new idea. Some of these were successful and retained on later engines, while others were discarded. In 1777 he built an engine using steam at a higher pressure, which was allowed to expand and so move the piston. It worked, but Watt preferred his low pressure engines and it was left to others to develop the high pressure *expansive* engine.

By the 1780s the Industrial Revolution was in full swing and factories needed power. Boulton realized this fact and persuaded Watt to produce an engine which could rotate a shaft in order to drive machines. The simple solution was to fit a crank, but a rival company had fitted one to a Newcomen engine and patented it. Rather than fight the patent Watt introduced an ingenious system of gears called a 'sun-and-planet', in which the 'planet' gear travelled around the 'sun'. Not only did this overcome the crank patent but also it drove the shaft to the factory at twice the speed of the relatively slow engine. The *rotative* engine was a great success, but it ran rather unevenly.

Watt decided to make both up and down strokes into working strokes. His first *double-acting* engine ran in 1783: it was smoother running and in addition it developed twice the power. An upward working stroke made it impossible to use a chain connection between the piston rod and beam, so Watt devised a parallel motion linkage based on the pantograph. This was one of his favourite inventions. Then came the conical pendulum governor to control the engine's speed, and by 1788 Watt's steam engine was complete and unrivalled for many years.

NATIONAL PORTRAIT GALLERY

Above: James Watt (1736–1819) has been called the greatest of British engineers.

Below: during his retirement, Watt worked in his garret workshop at Heathfield, Birmingham. Among other things, he became interested in making sculpturing machines. Below the centre bust on the table can be seen the original model of the separate condenser for steam engines.

SCIENCE MUSEUM

Trevithick's steam engine

Many of the early steam engines, such as those invented by Thomas Newcomen or James Watt, used steam at a very low pressure because most of their power was developed when the steam turned back into water. As the steam condensed it created a partial vacuum in a cylinder, and then atmospheric pressure (in Newcomen's engine) or low pressure steam (in Watt's design) moved the sliding piston on a working stroke. The idea of using high pressure steam to move a piston had been suggested as early as 1725 by Jacob Leupold of Leipzig but it is unlikely that his engine was ever built. In 1769 Watt took out his famous patent for the separate condenser which also included a clause covering the use of '. . . the expansive force of steam to press on the pistons . . .'. But Watt was opposed to the use of high pressure steam because of the danger of an explosion should a boiler or a pipe burst. He even suppressed the model made by his assistant William Murdock. In 1800 Watt's wide-ranging patent expired and the way was open for others to improve the steam engine.

Two of the leading inventors of this period were a Cornish engineer, Richard Trevithick, and an American, Oliver Evans; both advocated the use of high pressure steam despite much opposition. Trevithick had gained experience with a variety of steam pumping engines in the Cornish mines and for a while he worked for one of Watt's great rivals, Edward Bull. Then, in about 1798, he started experimenting with his own high pressure engine designs. The principal problem to be overcome was not the engine itself but a boiler to provide steam at pressures of perhaps 50 psi (3.5 bar). Trevithick developed a cylindrical boiler which could withstand these pressures and mounted his cylinder inside the boiler where it would be kept hot. The fire was also inside the boiler, carried by a circular flue which ran lengthways and back again in the shape of a U. This 'return flue' with a fire at one end and a chimney at the other was completely immersed in water to utilize as much of the heat from the fire and flue gases as possible.

One of the factors governing the power of a steam engine is the pressure acting on its piston. A Watt engine might develop a suction of 10 psi (0.7

RADIO TIMES HULTON PICTURE LIBRARY

Above: this view of Trevithick's steam car shows clearly how the cylinder, C, was located within the boiler, A.

Below: Trevithick's colliery locomotive. The largest wheel is a flywheel; the drive wheels are 3.75 ft (1.2 m) across.

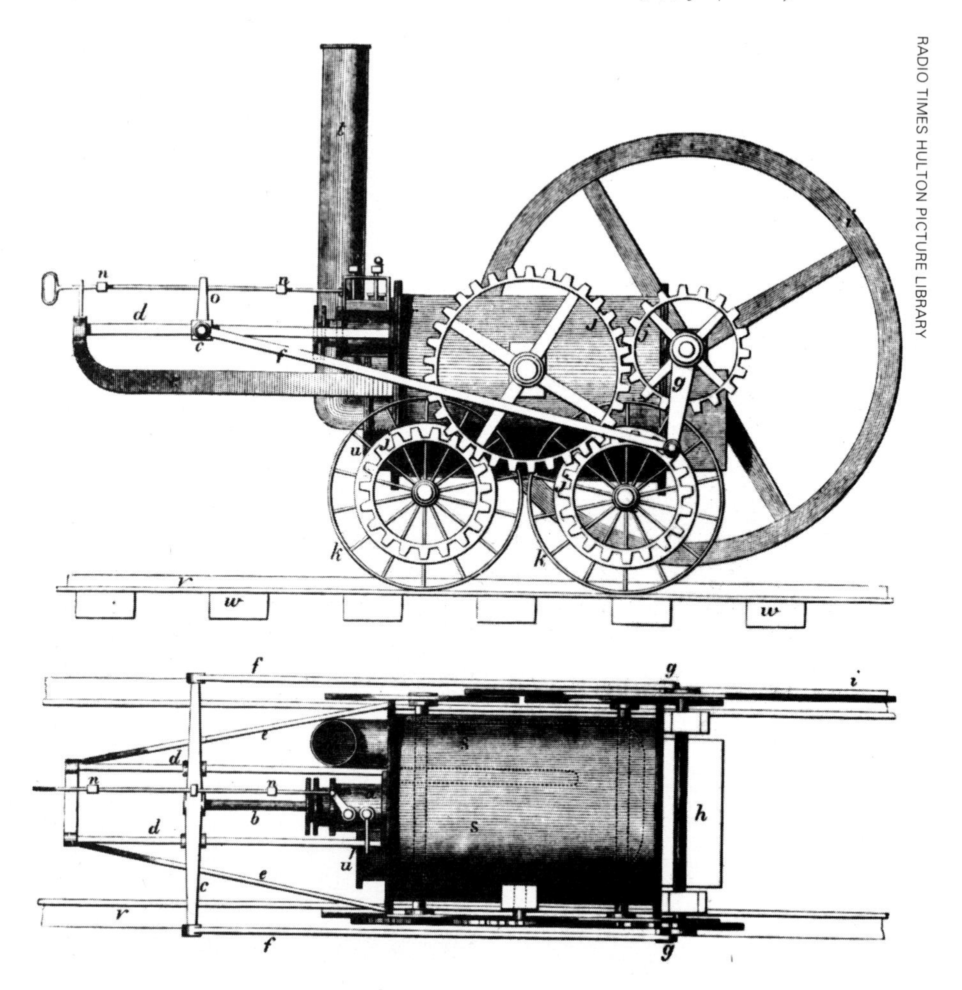

RADIO TIMES HULTON PICTURE LIBRARY

bar) on one side of its piston and a pressure of 5 psi (0.35 bar) on the other, giving a total differential pressure of 15 psi (1.05 bar) to move the piston. By comparison an engine working at 50 psi (3.5 bar) will develop more than three times the power or alternatively, the same horsepower could be produced by a smaller engine. This latter fact, coupled with Trevithick's compact boiler design and his elimination of the cumbersome see-saw beam, made the use of steam power for transport a feasible proposition at last.

In 1801, Trevithick, aided by his cousin Andrew Vivian, built a full size steam carriage and on Christmas Eve it was tried out on Beacon Hill, near Camborne in Cornwall. It lacked power but a few days later it was tried out on the turnpike with some success until a wheel caught in a gully and the coach overturned. Trevithick was not deterred and in March 1802 he obtained a patent for his high pressure engine and his steam carriage. A second steam carriage was built and demonstrated in London during 1803 but the roads were not really suitable and it too crashed. Trevithick suffered another setback in this year when the boiler of one of his stationary engines at Greenwich blew up. It was not a boiler fault however—a labourer had fastened down the safety valve while he had a meal.

The stationary high pressure engines continued to sell despite the accident and by September 1804 almost 50 had been built. One of these was an experimental engine built at Coalbrookdale to operate at 145 psi (10.2 bar), but 50 psi was the normal limit.

In 1804 Trevithick turned his attention to railways and produced his famous locomotive for the Pen-y-darren wagonway in South Wales. His second locomotive was built at Gateshead and has a doubtful history; then came the little *Catch-me-who-can* locomotive which gave joyrides near Euston Square, London in 1808. Trevithick returned to the design of stationary engines for the Cornish mines and in about 1812 he developed the Cornish boiler and the Cornish pumping engine.

MANSELL COLLECTION

Above right: 'Catch me who can' was on Trevithick's visiting card.

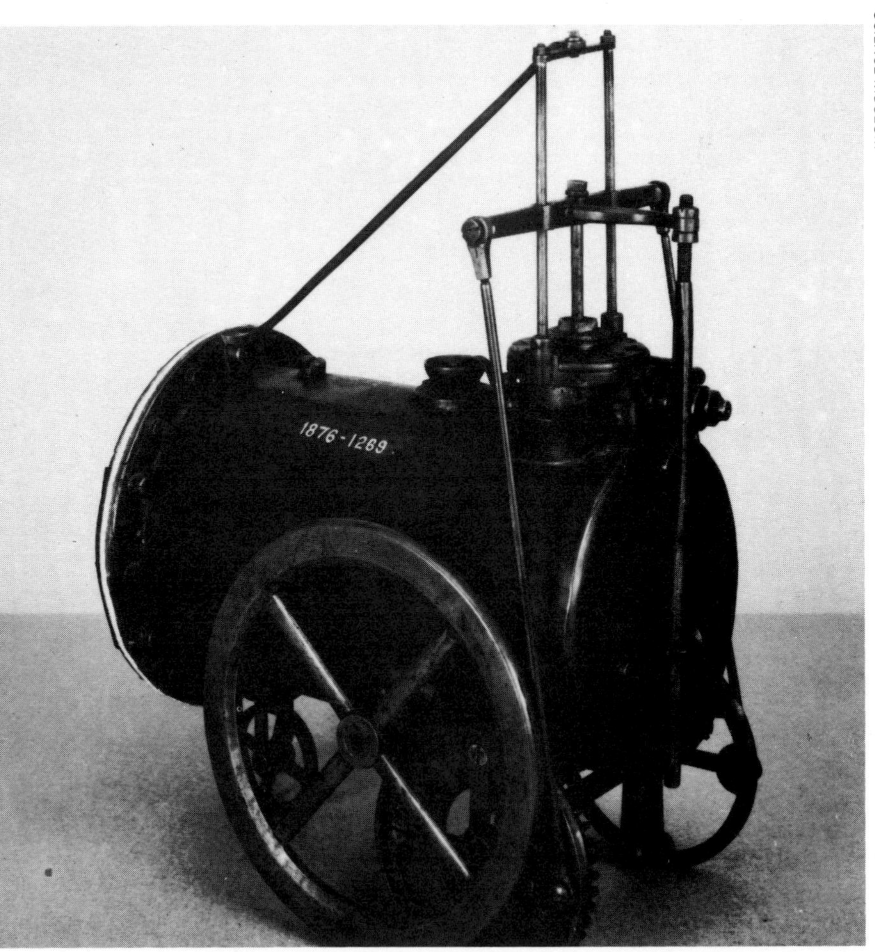

SCIENCE MUSEUM

Right: this model engine of Trevithick's, made in 1797, bears strong resemblance to 'Catch me who can' of 11 years later. The large wheel is a heavy flywheel.

The steamboat

The invention of the steamboat was not a clear-cut event and no one person could justifiably claim the honour of being the inventor. The claim is sometimes made for Henry Bell whose paddle steamer *Comet* carried fare-paying passengers on the River Clyde, Scotland, in 1812, yet before Bell there were over 50 other contenders. Some of the early suggestions were mere flights of fancy, but at the end of the 17th century the French inventor Denis Papin outlined plans for a boat with revolving paddle wheels driven by a simple steam engine. A few years later, in 1712, Newcomen produced his successful steam engine but this was not immediately adapted to power a boat: its great weight and low power output were a distinct disadvantage. In 1736 Jonathan Hulls of Gloucestershire patented a paddle tugboat powered by a Newcomen engine and although this boat was never built, a drawing survives.

From the mid 18th century there were many ideas and patents, not only covering the source of power, but also the method of propulsion, which included individual paddles, paddle wheels, screw propellers and even jets of water. But the first success came in 1783 when a paddle steamer called *Pyroscaphe* steamed for 15 minutes against the current of the River Saône in France. The designer of this vessel was the Marquis de Jouffroy d'Abbans, assisted by the Comte Charles M de Follenay. Jouffroy had earlier built a vessel powered by a two cylinder Newcomen-type engine but this had not been a success. The engine for *Pyroscaphe* was a much improved design with a single horizontal cylinder in which the piston developed power during each of its strokes. (James Watt also developed a 'double acting' engine which first ran in the same year of 1783.) Jouffroy made a model steamboat, which is preserved at the Musée de la Marine in Paris, but his work on full size steamboats was brought to a halt first by lack of money and then by the French Revolution.

Inventors in different parts of the world produced very different solutions to the same problem. In America, John Fitch built a vessel with a steam engine driving 12 paddles mounted vertically—six on each side of the vessel. This ingenious mechanism propelled the

Below: Jonathan Hull's patent for a paddleboat driven by a Newcomen engine included this drawing. The tugboat was never built, however.

Right: the 'Charlotte Dundas' showed that powering boats by steam was a practical proposition. It had a double acting horizontal Watt engine.

NAT MARITIME MUSEUM/MICHAEL HOLFORD LIBRARY

RADIO TIMES HULTON PICTURE LIBRARY

steamboat at about 3 mile/h (5 km/h) when it was tested on the River Delaware in 1786. Fitch went on to build several steamboats and one of them, called *Experiment*, was propelled by three separate paddles mounted at the stern. Fare-paying passengers were carried by the *Experiment* on the Delaware, but the service was not a financial success. In 1787 another American inventor demonstrated a very different type of boat, this time on the River Potomac. James Rumsey's boat was jet propelled, using a jet of water pumped out of the stern. Once again it worked but it was not a great success.

In Scotland, Patrick Miller carried out several experiments with paddle boats powered by men before turning his attention to steam. He enlisted the help of William Symington who had already gained considerable steam engine experience. Their first vessel was demonstrated on Dalswinton Loch near Dumfries in 1788. Its passengers included Robert Burns the poet, and Alexander Nasmyth, an artist, who painted a watercolour picture of the event. This shows a twin hulled vessel with two paddle wheels in tandem between the hulls. It was powered by a small two cylinder Newcomen-type engine which is preserved in the Science Museum, London. A second boat was built for use on the Forth and Clyde canal but abandoned when Miller withdrew his support. Symington eventually found another backer in Lord Dundas, who was a proprietor of the Forth and Clyde canal and favoured the idea of using steamboats to tow barges. The new boat was ready in 1801 and called *Charlotte Dundas* after the patron's daughter. It had a single paddle wheel let into the stern and this was driven by a double acting horizontal engine. *Charlotte Dundas* has often been described as 'the first practical steamboat' because in March 1803 she towed two loaded vessels along the canal for 19.5 miles (31.4 km). Unfortunately the other proprietors of the canal were worried about the damage caused by the wash from the paddle wheel and the project was abandoned. The *Charlotte Dundas* had been inspected by Henry Bell and an American, Robert Fulton.

After building an experimental steamboat in France, Fulton returned to the United States and produced the *Clermont*, with side paddles powered by a Boulton and Watt engine. In September 1803 the *Clermont* made the first of many successful voyages carrying fare-paying passengers on the Hudson River. At last the steamboat was a commercial success.

Below: Robert Fulton's 'Clermont' on the Hudson River in 1803.

Right: cross section of the 'Comet', built by Henry Bell and which carried passengers on the Clyde in 1812. The engine produced three horsepower. Steamships grew in size after this and by 1819 a steam-assisted sailing ship had crossed the Atlantic.

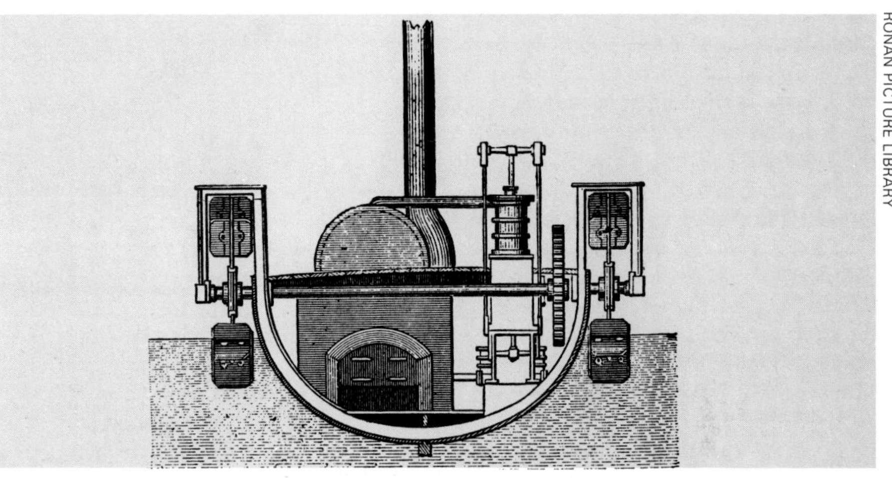

RONAN PICTURE LIBRARY

AMERICAN HISTORY PICTURE LIBRARY

The locomotive

At the start of the 19th century Britain was still undergoing the Industrial Revolution. Manufacturing industry, with its demands for raw materials and its output of mass-produced goods, was altering the pattern of life. Not only merchandise but people too needed to travel all over the country to help weave the web of commerce.

The appalling roads were totally unable to meet this need, while the canals were limited by the necessity of locks to overcome changes in level. Railways had already existed in Britain for about two centuries, the first examples being wooden trackways along which horses could haul waggons of coal from the mines to the nearest water transport—canal, sea or navigable river. In time the wooden rails changed to iron and the 'plateways' changed to edged rails along which flanged wheels were guided.

The early Newcomen and Watt steam engines were also adjuncts of the mining industry, but were stationary machines used to pump water or to haul waggons. The idea of placing a steam engine on wheels seemed highly irresponsible to the cautious Watt, who was well aware of the danger inherent in cast iron pressure vessels; but Richard Trevithick, a Cornish mining engineer, showed no such reservations. His lighter and more compact design of engine used higher pressure steam and proved far more adaptable. After building two steam road carriages, he turned to railway locomotives. The attention of the public at large was attracted by Trevithick's third railway engine, *Catch-me-who-can*, which gave demonstration rides in London.

The case for the commercial use of steam locomotives was given an unexpected boost by the Napoleonic wars, which increased the price of animal feed, making steam a more economic source of power than horses.

George Stephenson, the leading figure among the early locomotive builders was, like Trevithick, a mining engineer. His particular achievement was to synthesize the best available knowledge into a serviceable design that was to form the basis of all subsequent steam locomotives. By 1825, the most important year in the history of railways, Stephenson had achieved a commanding position.

MANSELL COLLECTION

The famous 'Catch-me-who-can' gave joyrides at Euston Square, London, for a shilling a ride, using Trevithick's one cylinder engine. This first ever railway was in operation for a month during 1808. The Euston terminus of the main line to the north today stands near this site.

The first public railway in the world, the Stockton and Darlington, was opened on 27 September 1825. In those early days the case for locomotive haulage was by no means proven and even as late as 1856 the S & D continued to use draught horses. On the hilly, western section of the line four inclines were worked by rope haulage and so on one railway the three main forms of motive power worked in conjunction. Nevertheless the locomotive was given pride of place, and the opening cavalcade of wagons was hauled by George Stephenson's *Locomotion*, preceded by a mandatory horseman bearing a red flag.

Although the locomotive proved relatively reliable and economical in service on the S & D, the battle was by no means won. When the directors of the first main line railway in the world, the Liverpool and Manchester, came to discuss the question of motive power they were not convinced of the suitability of steam. It was decided to hold trials in which various railway engines competed on the line at Rainhill for a prize of £500, though strenuous conditions had to be satisfied.

Of all the many entries at the start, by the day of the trial only three serious competitors remained. The first entry, Timothy Hackworth's *Sans Pareil*, represented the mainstream of locomotive design at that time. The requirement that all wheels should be sprung created problems, however, as the cylinders were placed vertically over the driving wheels and one cylinder even cracked during the trial itself.

The *Novelty*, entered by Braithwaite and Ericsson, was a four wheeled carriage with a unique vertical and horizontal boiler. Although ingenious, the *Novelty* developed insufficient power.

Thus the field lay open for the *Rocket*, entered by Robert Stephenson, George's son. Taking advantage of the rule that the load to be hauled was in proportion to the engine's weight, Stephenson designed a lightweight version of his recent locomotive, the *Lancashire Witch* of 1820. This engine proved a success. Enlarged and improved versions of the

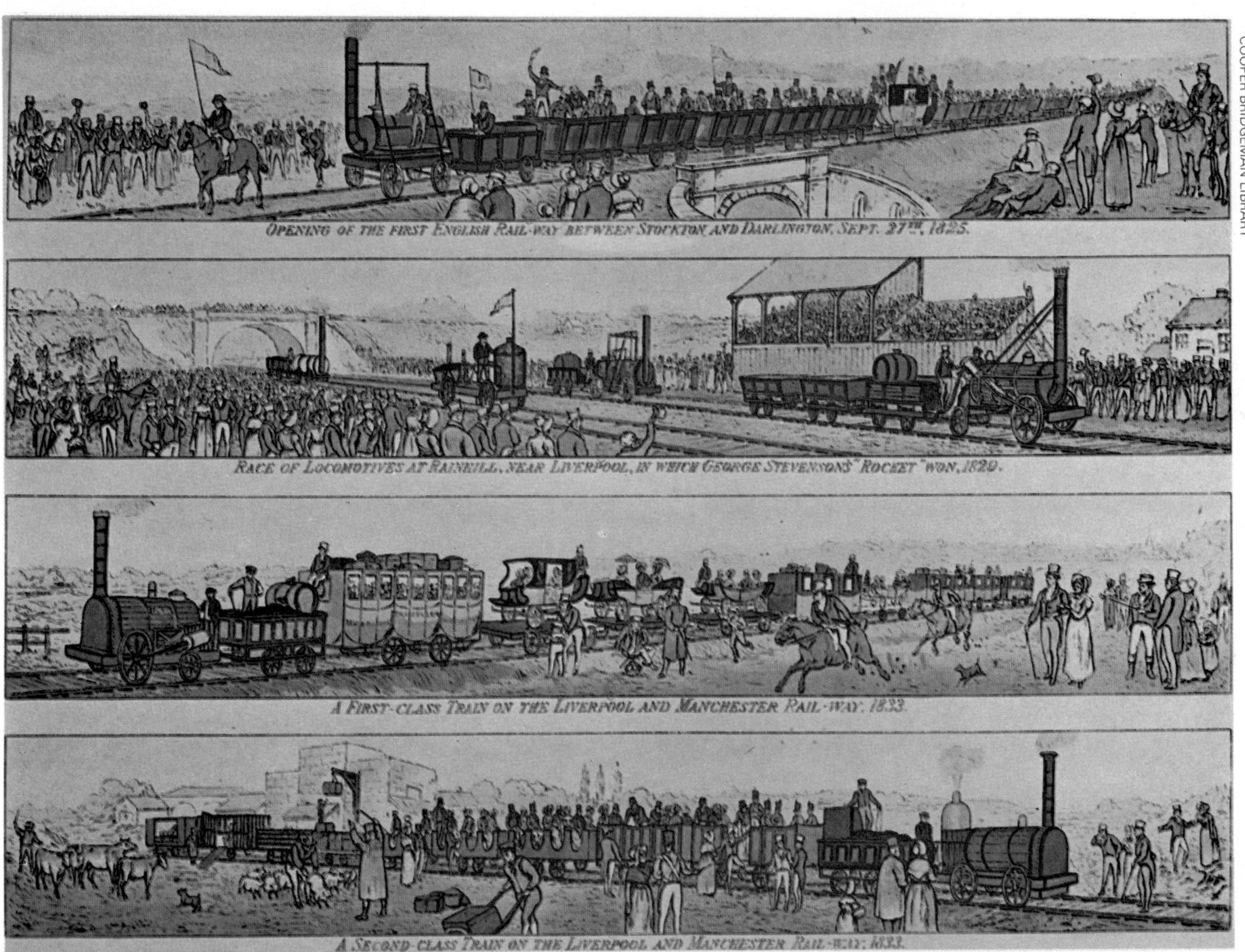

COOPER BRIDGEMAN LIBRARY

design became the standard motive power of the L & M, which was thus the first steam-worked passenger carrying main line.

The power and speed of the railway engine were quickly developed and soon a network of railway lines began to link the main towns and cities. Steam locomotives based on Stephenson's design were exported to many countries, including the United States, Russia, France, Belgium and Germany. British railway engineers too went overseas, building new railways in many parts of the world.

Top: scenes from railway history. Even though Stephenson had built the line, the Liverpool and Manchester's directors held the Rainhill Trials to decide whose locomotives should be used. Engines had to pull three times their own weight at a speed of 10 miles/hour (16 km/h), and had to weigh more than 4½ tons. The trials were won by Robert Stephenson's 'Rocket'; he then redesigned it to produce a class of engines with horizontal cylinders such as the Northumbrian (right).

THE NORTHUMBRIAN ENGINE.

SCIENCE MUSEUM

The cotton gin

The history of the 18th century revolution in the textile industry shows how, as new machines radically improved output in one branch of the trade, a need was straightaway felt for fresh inventions to enable other branches to catch up with demand for their products. By the 1780s, the whole cotton manufacture of England was leaping ahead—but now the supply of raw cotton could not match it. Lancashire still imported her vital raw material from India or the Middle East. A similar species grew in the American colonies, but in this species the short fibres clung much closer to the seeds than in the Old World types, and it was a hard job to remove them. So only a little was grown for domestic use.

By now the United States was an infant republic and many were convinced that progress would depend on finding some new staple agricultural product that would be better suited to American soils and climates than to those of the Old World, so that it could be exported to win capital for future development. The old staples on which the southern states' plantation economy had grown up, rice, tobacco and indigo, had all been hit hard in different ways by the War of Independence, and by the break with the British system which had nurtured them. Something was required which Europe (and particularly Britain) needed, but could not produce for itself. The desperate hunger of the booming English cotton industry suggested the answer.

Soon after the War, a long stapled, black seed variety was introduced from Barbados. It flourished along the coastal strip and offshore islands—hence the name of Sea Island cotton. But it would not grow in the vast reaches inland which were waiting for the new crop that could save them. Meanwhile the plantation system began to decline; hopes were expressed that slavery might just die of economic stagnation. One slave could only clean a pound (2.2 kg) of cotton a day by hand. An instrument like a roller mangle had been used in India for many centuries to separate the seeds. It worked fairly well with the long fibred Sea Island cotton, but when tried on the short fibred green-seed Upland type, it either broke the seeds if set close, or else let them through.

The solution to the problem was

Above: Eli Whitney's original cotton gin, showing the wire teeth which caught on the fibres and pulled them through.

Below: one of the many variations on Whitney's idea, using sawteeth to avoid infringing the basic patent details.

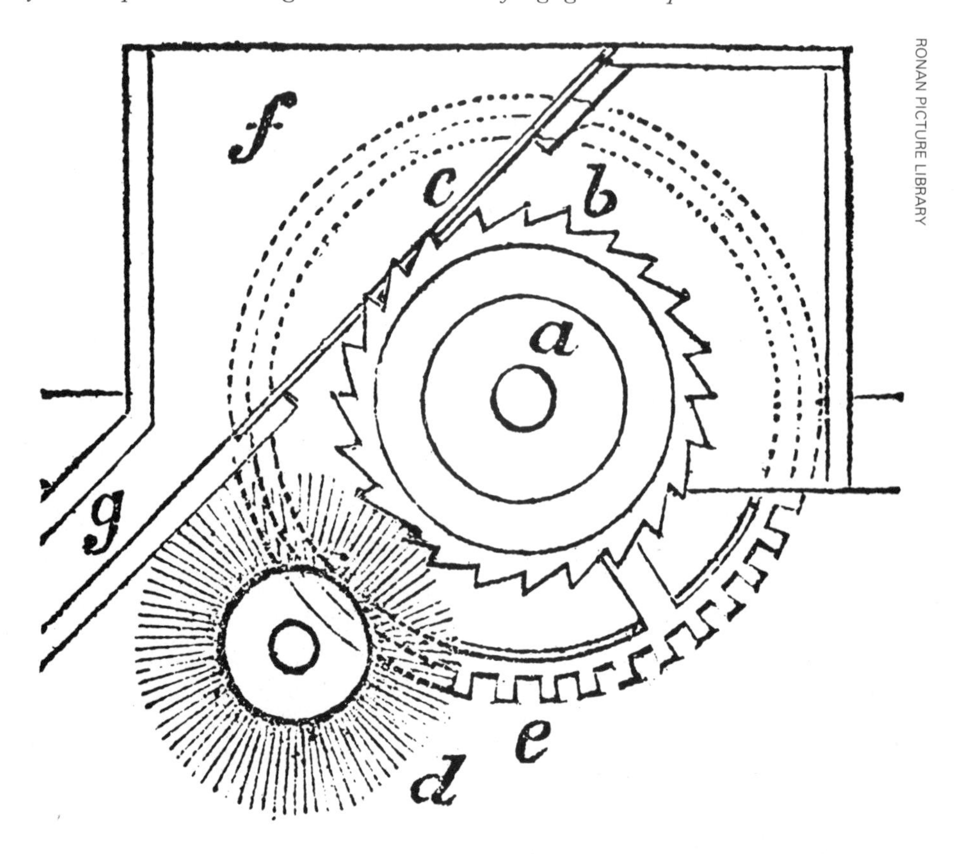

Whitney's cotton gin has an important place in American history, as this 1876 picture shows; the black slaves were probably less delighted than it makes out.

provided by a man who had no links with cotton, nor indeed was he a mechanic by trade. The son of a Massachusetts farmer, Eli Whitney showed while a boy that he was good with his hands, and quick at mathematics; and good at business too, for at fourteen he had started his own business, manufacturing nails. When his market was flooded at the return of peace and British goods, he turned to hat-pins. He also had hopes of rising into the professional class, and saw himself through a local academy and into Yale University by teaching school. Graduating in law, he planned to establish himself first as a teacher, but the only post offered him was as private tutor to the children of a wealthy Georgia planter. While travelling to Georgia, however, he met Phineas Miller, administrator of the estate of the widow of Nathanael Greene, one of the leading generals in the Revolutionary army. Whitney impressed both Miller and Mrs Greene; they encouraged him, and invited him to stay on the plantation. So this chance gave Whitney entry into a community whose apparent luxury and freedom of manners were very exciting to a New England Yankee —a community which was waiting upon a quick and easy method of separating cotton from its seeds. 'If a machine could be invented which would clean the cotton with expedition it would be a great thing both to the Country and to the inventor', he wrote to his father. Quite rapidly he formed an idea in his mind, and began to develop it. During the winter of 1792 he worked out the details, and by the spring had constructed a satisfactory model. Miller offered him financial support and they went into partnership.

Whitney's machine, like the old one, had two rollers. But now they were equipped with means to pull at the cotton lint. The main roller was fitted with rows of wire teeth 'set in annular rows'. He had thought of mounting rings of toothed sheet iron, like circular saws, upon the cylinder, but could not get hold of any. One of the Greene children had bought some iron wire to make a bird cage, 'and seeing this wire hanging in the parlour it struck me I could make teeth with that'.

The cotton was fed into a hopper whose base was a metal plate he called the *breastwork*, in which slots were cut. Underneath it was placed the cylinder, so that as it turned the teeth passed through the slots and drew the cotton down. But the seeds were too large to go through, and the gap between the teeth was also 'so small as not to admit a seed nor half a seed to enter it'. The cotton would have then wrapped itself round the cylinder, but a second roller called the *clearer*, driven from the first by means of a belt, carried rows of little brushes which forced it off and out, ready for collection. Nothing was needed but to turn a crank and the first roller began to pull the cotton down. And a crank could easily be adapted so as to be driven by a horse, water—or steam.

Whitney's *gin* (or engine) was an immediate success, and he set up a factory to go into production. Indeed, it was too successful and many potential users refused to pay the lavish royalties in kind on which Whitney and Miller insisted. Instead, other similar machines appeared, working on the same lines but with just enough differences to make it necessary to go to law to prove the point. Whitney became embroiled in legal battles, and spent more than he gained in the process.

The other effects of his invention had a significant part to play in American history. More and more land was taken from the Indians and put under cotton, and slavery became highly profitable again. The southern planters were confident that the Union depended on them and on the wealth they were bringing into the country. This explosive state of affairs eventually led to the American Civil War.

MODERN TIMES

In the 1936 film *Modern Times*, the 'little fellow' portrayed by Charlie Chaplin scratched his head at the complexity of contemporary life. Many years before that, someone had already suggested that the Patent Office be closed, since everything had already been invented. Yet today the flow of new ideas is faster than ever. There is even some danger that we will invent more technology than we need. The vitality of global Capitalism and the exploration of Space provide new motives.

The internal combustion engine

The origins of the internal combustion engine can be traced back to the late 17th century, when Christiaan Huygens and Denis Papin experimented with an engine in which gunpowder was exploded inside a cylinder to move a piston. Nearly two centuries elapsed before a working internal combustion engine emerged, a period in which the steam engine reigned supreme.

It was in 1860 that a French engineer, Etienne Lenoir, produced the first successful internal combustion engine. This resembled a horizontal double acting steam engine but instead of using steam to move its piston, a mixture of gas, such as coal gas, and air was ignited inside the cylinder by a spark. Lenoir's gas engine was an immediate success as a small industrial power source, for despite its high gas consumption it was compact and convenient to operate.

Several improvements to increase the efficiency of the gas engine followed, including a version operating on the atmospheric principle used earlier by Newcomen's steam engine. The most notable development was the introduction of a compression stroke to increase the pressure of the mixture of gas and air before it was ignited. Credit for the invention of this 'four stroke cycle' is usually given to a German engineer, Nikolaus Otto, of the firm Otto and Langen, although it had been suggested earlier by Alphonse Beau de Rochas in Paris. Gas engines using the 'Otto cycle' were made from 1876 onwards and their improved performance led to a rapid increase in popularity as a stationary source of power. Their ability to provide instant power without the inconvenience of a boiler resulted in many small steam engines being superseded. Gas engines were not, however, very suitable for powering vehicles because they were slow running and they needed a supply of gas.

Petrol [gasoline] vaporizes very easily, and it became clear that petrol vapour could replace gas. In 1881 an employee of the firm Otto and Langen left to produce his own, high speed petrol engine. Gottlieb Daimler, assisted by his close friend Wilhelm Maybach, worked long hours and by 1885 they had an engine working and patented. It had a single vertical cylinder which developed ½ horsepower using the four stroke cycle but, more important, it was powered by liquid fuel and it rotated at about 700 revolutions per minute. This was very fast by gas engine standards. Daimler offered the engine to his former employers but they turned it down. Undeterred, Daimler fitted his engine into a primitive motor cycle.

In the same year, 1885, another German engineer, Karl Benz, working independently, produced a three wheeled car with an internal combustion engine but the Benz engine was not as advanced as Daimler's: it was heavy and slow running. The following year Daimler produced a four wheeled car with a high speed petrol engine and this car can be claimed to be the forerunner of the modern automobile.

The significance of the invention of the petrol engined car was not immediately apparent because the railways were a well established form of transport by the late 19th century, whereas cars were expensive and unreliable. In addition the internal combustion engine was not the only source of power available for cars because steam engines and electric motors were also being used success-

Below left: Lenoir's first gas engine of 1860. This was inefficient, as the gas was not compressed before ignition. The spark was produced by the ignition coil R from the batteries on the left, just as in the system of a modern car.

Below right: a version of the Otto and Langen gas engine with its cylinder inside the ornamented pedestal.

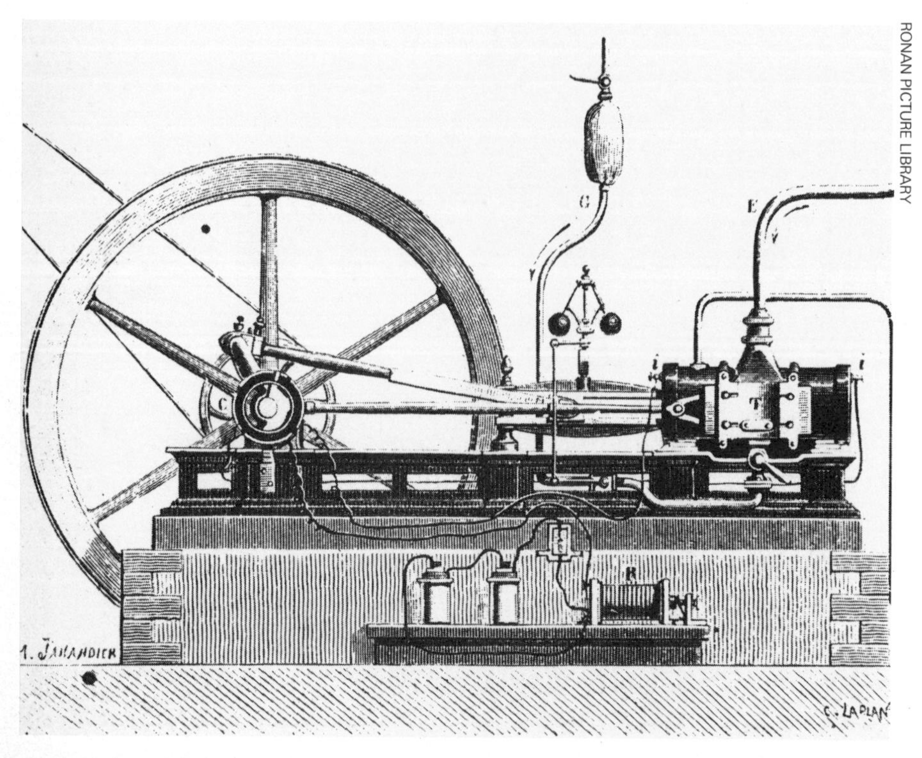

RONAN PICTURE LIBRARY

RONAN PICTURE LIBRARY

ROYAL SCOTTISH MUSEUM

ROYAL SCOTTISH MUSEUM

fully. The mass production methods introduced in America by Henry Ford, during the early years of the 20th century, played a major role in establishing the petrol engine in the dominating position it holds today.

While the petrol engine was being introduced, another type of internal combustion engine emerged as a practicable power source. This was the oil engine, which made use of low grade fuel oil, and in its early days closely resembled the petrol engine. One of the first examples to become a commercial success was introduced in 1888 by the firm of Priestman Brothers of Hull, England. The fuel and air mixture was compressed and heated before entering the cylinder where it was ignited by an electric spark. Later engines operated without a spark but retained some form of heating.

In 1892 Rudolf Diesel in Germany patented a compression-ignition engine which dispensed with the heating device altogether. The air in the cylinder was compressed by the piston to a very high pressure, then the fuel was injected and the high temperature resulting from the compression caused it to ignite spontaneously. This engine, now known as the diesel, became widely used.

Just as the reciprocating steam engine was followed by the steam turbine, so the reciprocating internal combustion engine was followed by the internal combustion turbine—the gas turbine. After early trials in 1905, the gas turbine finally emerged as a successful engine during World War 2. In recent years another type of internal combustion engine has emerged—the Wankel rotary engine with its triangular shaped rotor instead of reciprocating pistons, but still using the four stroke cycle.

Above left: Daimler's motor cycle of 1886 was a very solidly constructed affair with wooden wheels, the rear of which was driven by a leather band from the petrol engine. Ignition was by means of a heated tube. To support the heavy machine when stationary, the rider lowered two side wheels with his feet. The machine was only experimental and may never even have run very successfully.

Left: the Rolls Royce Derwent of 1945, used in the Gloster Meteor. In this form of internal combustion engine, hot gases from the central combustion chamber drive a turbine to run a compressor then emerge from the jet on the left.

The telegraph & telephone

No single person invented the telegraph. For over a hundred years before the first true telegraph messages were transmitted, various people had shown that the idea of sending electric impulses along a wire was a practical one. In 1729, Stephen Gray of Charterhouse in London found that a static electric charge could be passed 700 feet (229 m) along a conducting thread. A few private experiments using static electricity were carried out in Spain and England, but it was not until the discovery of current, rather than static, electricity by Volta in 1800 that experiments with electromagnetism could begin. Within a few years various systems for sending messages over a distance had been tried.

It was in 1836 that William Cooke, a young Englishman recently invalided out of the East India Army, saw a demonstration of such a system devised by a Russian diplomat, Baron Schilling, in Heidelberg. On returning to England, Cooke made several similar devices. In the course of his experiments, he found that Professor Charles Wheatstone of King's College was engaged in similar work, and the two teamed up.

Unlike some other devices then being tried which had a separate wire for each letter of the alphabet, Cooke and Wheatstone's first device had just six wires operating five needles. Electrical signals sent through the wires to coils caused deflections of the needles; the needles were in a line at the centre of a grid of letters so that moving any two indicated a particular letter.

The Liverpool and Manchester Railway, and in particular Robert Stephenson the engineer, showed interest in the scheme, but despite a successful demonstration alongside the railway at Euston in London they did not adopt the system. Cooke then managed to involve Brunel, builder of the Great Western Railway, who agreed to a trial between Paddington in London and West Drayton, some 13 miles (21 km) along the line. This was set up in 1839, and was used at first for reporting on the positions of trains with great success. At the time, railway signalling was almost non-existent, relying on the vigilance of drivers to check that there were no other trains in the way.

In 1843, the service was extended to Slough, a further five miles (8 km), using a two needle system in which the particular letter out of several was indicated by a code of needle movements.

The Cooke and Wheatstone telegraph was made available for public use, and thus represents both the first commercial telegraph service and the first commercial use of electricity. But at about the same time in the USA, Samuel Morse was trying to convince Congress to allocate him funds to develop his own system—which in time was to become the most widely used and effective.

After a close vote, with many abstentions from people who did not understand what it was all about, he got his money and in 1844 the first messages were sent, using Morse's code, over the 40 miles (64 km) between Baltimore and Washington. The Morse system, requiring only one wire, had many

Below left: one of Cooke and Wheatstone's original needle terminals. The service created great interest (below); the speed of signalling was overestimated, the true value being close to 186,000 miles/sec.

SCIENCE MUSEUM

SCIENCE MUSEUM

THE WONDER of the AGE ! !
INSTANTANEOUS COMMUNICATION.

Under the special Patronage of Her Majesty & H.R.H. Prince Albert.

THE GALVANIC AND ELECTRO-MAGNETIC
TELEGRAPHS,
ON THE
GT. WESTERN RAILWAY.

May be seen in constant operation, daily, (Sundays excepted) from 9 till 8, at the
TELEGRAPH OFFICE, LONDON TERMINUS, PADDINGTON AND TELEGRAPH COTTAGE, SLOUGH STATION.

An Exhibition admitted by its numerous Visitors to be the most interesting and ATTRACTIVE of any in this great Metropolis. In the list of visitors are the illustrious names of several of the Crowned Heads of Europe, and nearly the whole of the Nobility of England.

"This Exhibition, which has so much excited Public attention of late, is well worthy a visit from all who love to see the wonders of science."—MORNING POST.

The Electric Telegraph is unlimited in the nature and extent of its communications; by its extraordinary agency a person in London could converse with another at New York, or at any other place however distant, as easily and nearly as rapidly as if both parties were in the same room. Questions proposed by Visitors will be asked by means of this Apparatus, and answers thereto will instantaneously be returned by a person 20 Miles off, who will also, at their request, ring a bell or fire a cannon, in an incredibly short space of time, after the signal for his doing so has been given.

The Electric Fluid travels at the rate of 280,000 Miles per Second.

By its powerful agency Murderers have been apprehended, (as in the late case of Tawell,)—Thieves detected; and lastly, which is of no little importance, the timely assistance of Medical aid has been procured in cases which otherwise would have proved fatal.

The great national importance of this wonderful invention is so well known that any further allusion here to its merits would be superfluous.

N.B. Despatches sent to and fro with the most confiding secrecy. Messengers in constant attendance, so that communications received by Telegraph, would be forwarded, if required, to any part of London, Windsor, Eton, &c.

ADMISSION ONE SHILLING.

T. HOME, *Licensee.*

Nurton, Printer, 48, Church St. Portman Market.

advantages over the more complex Cooke and Wheatstone machines. In the USA the use of his telegraph spread rapidly. Gradually other countries began to adopt the single wire and code system, and by 1852, just nine years after the first messages were sent from Paddington, there were some 40,000 miles (64,000 km) of telegraph lines in operation throughout the world. Ten years after that, British ministers could telegraph their subordinates in India, and telegraph wires had spanned the American continent, putting the Pony Express out of business.

Once the use of the telegraph system was made public, its possibilities were soon exploited. Not only could railway accidents be avoided, but newspapers could report more up to date news, businessmen could transact deals with people hundreds of miles away, farmers could get a better price for their wheat, distant relatives could keep in touch and isolated communities could become a part of their nation. It was the telegraph, rather than the telephone, which brought about a revolution in means of communication.

If the telegraph expanded rapidly after its beginnings, the telephone was rather slower off the mark. It was on 10 March 1876 that Alexander Graham Bell sent the first telephone message—on spilling some battery acid over himself he called out 'Mr Watson, come here—I want you,' through the microphone. Watson, on another floor, heard his voice through the telephone. Although lines were soon set up in the USA, the use of the telephone was largely limited to comparatively short distance calls, telegrams being sent over larger distances. Even so, by 1885 when the New York to Boston telephone line was laid, there were nearly 70,000 telephone subscribers in the world.

By 1927, Britain's telephone system had only 6000 miles of long distance cable, but in that year the first transatlantic link was established by radio. It was not until 1956 that a transatlantic cable for telephone use was laid—only six years before regular communication by satellite began.

Above left: Bell's first telephone, which first transmitted sounds on 3 June 1875.

Left: an 1879 switchboard for 50 lines. Because Strowger, an undertaker, lost business to other firms through operator errors, he invented automatic switching.

Photography

The first photograph ever taken was a view of rooftops, recorded in 1826. Its creator was Nicéphore Niépce, a Frenchman, and its materials were quite different from those of today: the image consisted of areas of asphalt hardened by the action of light. Furthermore, its eight hour exposure time was most impractical.

Although this first photograph was a technical dead-end, it was Niépce's partner, Daguerre, who tried using silver iodide as the light sensitive material. At first, his attempts produced no results, and Daguerre put his trial plates away in a cupboard to await repolishing. When he returned a few days later, he found that what turned out to be mercury vapour had made the images exposed on the plates visible. In 1839, the new Daguerrotype process was released, and photography was a commercial proposition.

Daguerrotypes are delicate, rather difficult to view and laborious to prepare. At about the same time as Daguerre was making his experiments, William Fox Talbot was discovering a different system—the one which, in principle, is the basis of modern photography.

Fox Talbot used paper impregnated with light sensitive materials such as silver chloride or, later, silver nitrate. He discovered a way to 'fix' the image, so that it would not be affected by further exposure to light. On the suggestion of Sir John Herschel, who had invented a

SCIENCE MUSEUM

Above: Roger Fenton, who took his bulky photographic apparatus to photograph the Crimean War of 1854 to 1856.

Below: these pictures are the negative and positive of a Calotype, taken by Fox Talbot himself. The positive shows more graininess because it had to be printed through the paper of the negative. This was avoided by using glass plates.

ROYAL PHOTOGRAPHIC SOCIETY

similar process, in 1839 he used sodium thiosulphate as an improved fixer—still very widely used. Fox Talbot's images were on paper negatives, which could then be used to make positive prints for viewing. This meant that multiple copies of a photograph could be made, and Fox Talbot brought out the first book illustrated with photographic prints, *The Pencil of Nature*, in 1844.

Fox Talbot's Calotypes still required exposure times of the order of minutes, and the prints suffered the disadvantage that they showed the irregularities of density in the paper negatives, through which light had to pass to make the prints. The answer was to coat the light sensitive materials on to glass; but since glass cannot absorb the chemicals, some form of carrier was needed.

The first successful glass plates were made by Frederick Scott Archer in 1851. The secret of his process was the recently-discovered *collodion*, a viscous, filmy cellulose solution made by nitrating cotton. The silver halide grains could be dissolved in a collodion layer on a glass plate. Despite the disadvantage that the plates had to be coated on the spot and used while still wet, this process was responsible for a great increase in interest in photography. Photographic studios and travelling photographers became common, but because of the need to coat plates with chemicals, photography was limited to a relatively small number of practitioners.

The advance which released photographers from the chore of coating their own plates was suggested by an amateur photographer, Dr R L Maddox, in 1871. Instead of collodion, he suggested using *gelatin* to carry the silver grains. Gelatin, a jelly-like substance made from the hooves and hides of cattle, has the property that although it dries to a hard layer, it swells when wet. The developing agents can thus reach the silver grains, whereas collodion does not re-absorb water once dried. It was found that the use of gelatin could also make plates much more sensitive.

Photographic manufacturers now could go into business coating glass plates for anyone with a camera to use, resulting in a great increase in the number of photographs taken, and an improvement in coating quality. One of these manufacturers, an American called George Eastman, realized that even pre-coated glass plates were unnecessary. His idea was to coat the gelatin on to a flexible support. At first this support was paper, from which the gelatin had to be stripped, but in 1889 Eastman introduced nitrocellulose film. Eastman's Kodak (an invented word) box camera contained enough film for 100 photographs; the whole thing was returned to the factory for processing. Flexible film also made movie photography possible.

Photography now became practical for everyone, and the concept of a simple camera with few controls other than the shutter has remained popular: the mass market has always supported the trade, and without it many improvements in photography would never have been made. An example is the introduction of Kodachrome in 1935. Two New York music students. Leo Godowsky and Leopold Mannes, also keen amateur photographers, realized the potential of a colour film that would be as easy for the amateur to use as black and white film. They studied part-time, supporting themselves by their musical ability, carrying out their experiments in Mannes' kitchen. Eventually they joined a Kodak team, and the result, Kodachrome, with subsequent improvements, is still the most popular reversal film.

Some of the first colour photographs to be taken were particularly beautiful. This one was taken in 1908 by Heinrich Kühn on Autochrome film, invented by the Lumière brothers, but difficult to use.

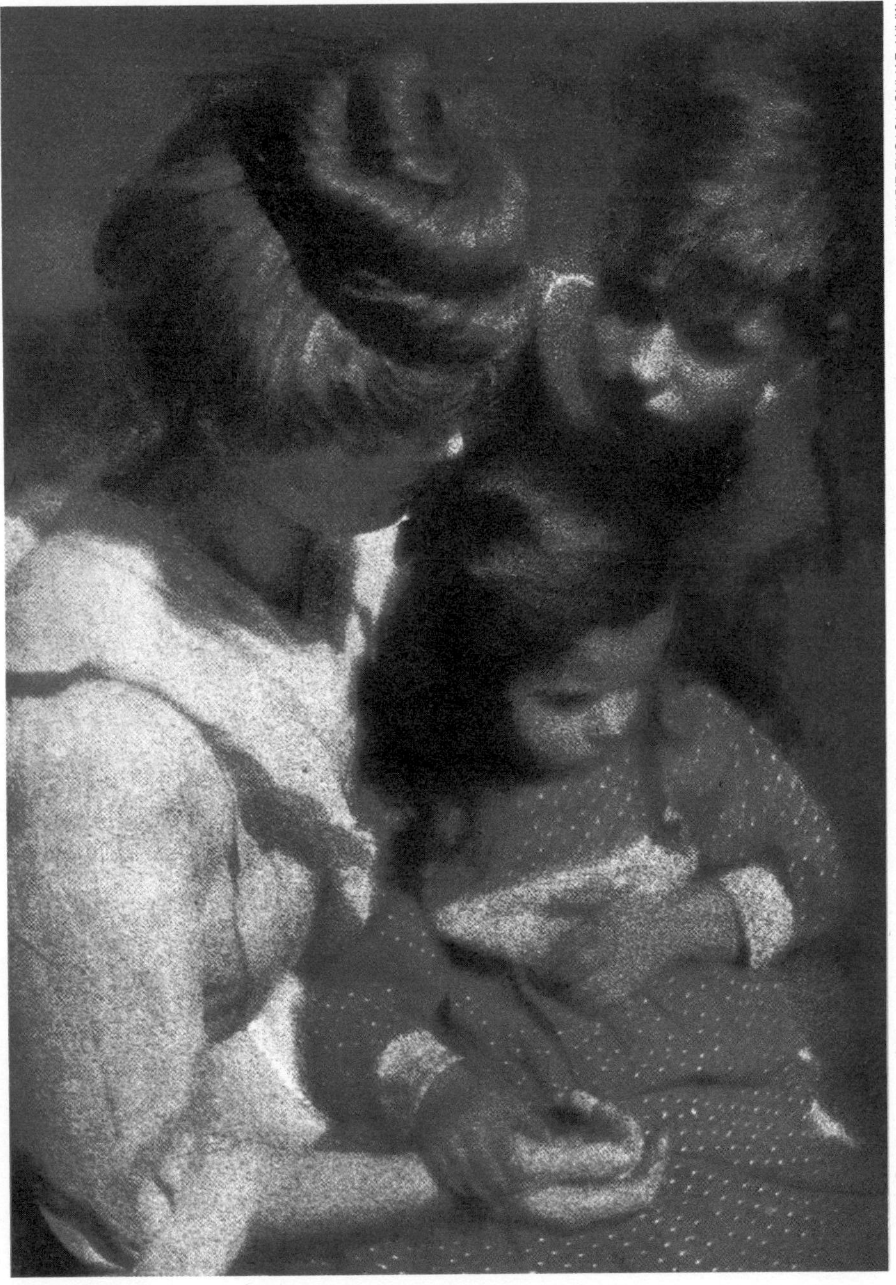

MARY EVANS PICTURE LIBRARY

Electricity generation

The science of electricity as we know it today—current rather than static electricity—began in 1800 with the invention of the electric battery by Volta. Throughout the 19th century, new discoveries were made about what was originally thought of as the 'electric fluid', and one hundred years after Volta's invention, electricity was being supplied and used in offices, public places and homes.

In the same year, 1820, that Oersted announced his discovery of a magnetic field surrounding a wire carrying electric current, Arago made the first electromagnet; and a year later, Faraday showed that the effect would work the other way round—that is, moving a magnet close to a wire produces an electric current. This is the basis of all electric generators.

In practice, the wire is arranged in a coil which is rotated between the poles of a magnet. At first permanent magnets were used, then electromagnets operated by separate batteries. Finally it was realized by Charles Wheatstone that the current from the generator could run its own electromagnet: there is always a slight residual magnetism in an electromagnet which will start the generator off.

The encouragement to develop the means for generating electricity did not, at first, come from the need to drive machinery or provide heat. The first generating companies were set up solely to run electric lights.

Even at the beginning of the 19th century, one finds references to 'electric light'. This light was not, however, that produced by the now-familiar light bulb, but that of the *carbon arc* or, originally, simply the sparks produced by static electricity. The brilliant light from a pair of carbon rods connected to a supply of about 50 volts far exceeded that which could be produced by its main rival, the gas lamp, which had been introduced in the early years of the 19th century.

MARY EVANS PICTURE LIBRARY

Top right: an Edison-installed electric plant in Paris in 1887. Each dynamo was capable of producing enough current to light either 500 or 1000 lamps, working on the direct current system.

COOPER-BRIDGEMAN LIBRARY

Right: at right is Edison's first light bulb, with carbon filament; on the left, the first commercial light bulb.

The carbon arc, however, needed constant attention and was therefore limited in use. The first important installations were in lighthouses round the coasts of Britain: that at South Foreland, the first, was installed in December 1858. Run by a steam engine, the rather inefficient generator produced less than 1½ kW (for comparison, the average electric fire consumes one kilowatt of power). Nevertheless, the light was considered of good quality and turned out to be not much more expensive than the oil lamps then in use, in terms of cost per light unit.

The use of arc lamps was limited to a few specialized applications until a lamp that did not need constant attention was invented. In addition, the early generators were rather poor. But with the invention in France of an improved generator by Zénobe Théophile Gramme, and of what was called an 'electric candle'—a self operating arc lamp—by Paul Jablochkoff, during the 1870s, electric lighting became increasingly common in large city shops, factories and public places such as stations.

Even so, electric lighting was not on a scale where it could be enjoyed in the home, and the brilliant, noisy, smoky arc lamps were quite unsuitable for small rooms. The invention which made all the difference was the incandescent filament lamp, working on the same principle as today's light bulb. It had been known for some time that a current applied to a thin conductor, such as platinum wire or carbon strip, would cause it to glow brightly, and that it would quickly burn up if it were not enclosed in a vacuum. But the early lamps, made in the middle of the century, had a very short life as a result of the poor vacuum pumps of the time which left a considerable amount of air in the bulb. When Joseph Swan in England and Thomas Edison in America both approached the problem independently in the late 1870s, they had the advantage of improved vacuum pumps. To avoid patent problems, they joined forces in 1883; a year earlier, Edison had started the first commercial electric light companies in both America and Britain.

The growth of the electric light industry is revealed by the numbers of lamps which Edison's generators ran. As soon as Edison's Pearl Street plant opened in 1882, it supplied 2323 lamps. By 1884, it was supplying 11,272 lamps in 500 premises, while the individual generators which he supplied were lighting nearly 60,000 more. By 1885 there were some quarter million lamps in use in the United States. Similar growth was seen in Britain, as people became familiar with the advantages of the new lighting: by the end of the century there were 2½ million lamps in London alone. The power supplied for lighting was soon put to use, notably in factories at first, where the development of enclosed motors meant that motive power could be provided wherever it was needed.

MANSELL COLLECTION

MANSELL COLLECTION

London's first main power station was at Deptford, on the Thames (above left), and was in the charge of Sebastian de Ferranti. He introduced the idea of generating the electricity at high voltage and using transformers to reduce this locally. His dynamos (left) were 42 feet (14 m) across. A report of 1889 said 'The glow lamps should be distributed over one's house in the most advantageous positions, and they should be "switched" off when not required.'

Sound recording

The most surprising fact concerning sound recording is that it should have taken so long for the simple formula 'sound produces vibration, therefore vibration produces sound' to be given practical application. The very simple mechanical device involved could have been manufactured centuries before by craftsmen who made precision instruments such as clocks. These had been spring driven from around 1450. And even in the early cultures, the Chinese and Egyptians were aware that sound produced vibrations. Scientists and writers ever since then had dreamed of recording and reproducing machines without understanding the simple secret of their operation. Cyrano de Bergerac, writing in 1649, talks of a box 'full of little Springs and imperceptible Engines' from which, when it was wound up, man could hear sounds 'as from the Mouth of Man or a Musical Instrument'.

Electric recording might easily have preceded acoustical recording, for practical attempts to transmit speech over electric circuits were being made as early as 1854, though the aim was communication by telephone, not the recording of sound. Alexander Bell, who patented telephony in 1876, was not thinking in terms of entertainment; nor, curiously, was Thomas A Edison whose mind was also firmly fixed on communication. Many men were just about to hit on the truth but it turned out that Edison was to be the figurehead.

He saw the link between the physical impact that a sound could make and its reproduction back into sound, and concentrated his voice down a primitive horn so that it vibrated a diaphragm. If the diaphragm vibrated so would a stylus or needle of some sort connected to it. And if the stylus cut a groove in some soft material on a revolving cylinder the traces of the vibrations would be recorded. Reverse the operation so that the groove vibrated the stylus and hence the diaphragm, and then the original sound should come out of the horn.

This he tried with a simple machine built by his colleague John Kreusi, using tinfoil as his soft material, and had the experience of hearing a repro-

COOPER-BRIDGEMAN LIBRARY

Right: the stylus of Edison's phonograph was first fixed to cut the groove, and was then allowed to vibrate to cut the recording, guided by the screw thread.

Below: Edison's perfected phonograph in use in London's Crystal Palace in 1888.

RADIO TIMES HULTON PICTURE LIBRARY

duction of his own voice reciting 'Mary had a little lamb'. He patented his *phonograph* on 19 February 1878.

Edison eventually began to produce cylinder phonographs, improving their mechanism and using wax as the material for the cylinders. The cylinder was to remain in production until 1916, but meanwhile other competitors, notably Emile Berliner who invented the flat disc recording in 1887, were exploiting his ideas. At first the gramophone was still thought of as something useful in commerce and it took some time to realize that its major realm was to be in entertainment. The first commercial records and machines were not produced until 1889 in Germany and not until 1893 in America.

Sound recording really took flight in the 20th century. Magnetic wire recording had been invented in 1898 but the commercial recordings stayed with acoustical principles until 1925, when the first electrical recordings were issued by Victor and Columbia in America. In electric recording, microphones and amplifiers are used, rather than the vibration of the sound alone. High fidelity had been bandied about as a salesman's motto, but true full frequency range recording was not achieved until 1944. The microgroove LP recording was introduced in 1948 (although it had been a possibility long before) and stereo discs were generally introduced in 1958.

With the increasing improvements of the recording process, musicians became interested in making recordings for their own sake; nowadays, some classical artistes never give concert performances but are known solely by their recordings, notably Glenn Gould, the pianist. At the same time, high quality recordings have brought music to a wider audience than ever before which has not led to any decrease in interest in live performances.

Tape recording was first perfected in Germany during World War 2, and during the 50s and 60s became a popular domestic hobby. Before the mid 1950s, broadcast items all had to be recorded on disc if they were wanted again, even by large broadcasting companies. Tape recording was first used to improve the production of his radio show by Bing Crosby, and has spread to become an invaluable aid to the making of discs by multiple track recording and repeating of sections to correct errors. Multi-tracking was popularized in the mid 50s by Les Paul, who recorded guitar and voice songs with his wife, and by the early 60s it was a standard technique. Nowadays, some pieces of music can exist only as recordings, since the performer records himself playing a large number of individual instruments.

The invention in 1963 of the Compact Cassette brought about a revolution in domestic tape recording. The comparatively low cost, convenience and standard format of the apparatus means that anyone may record sounds at any time: it is comparable to the introduction of film in the late 19th century, which replaced the bulky plate cameras.

DEUTCHES MUSEUM

WONDERFUL.

THE 20TH CENTURY has already produced many marvellous inventions and improvements, but there is practically nothing in which such astounding strides have been made as the

TALKING MACHINE ART.

ONE OF THE GREATEST disadvantages of the Phonograph or Talking Machine has hitherto been its great cost, but after many years of experiments we are at last in a position to offer our greatest friends—viz., THE BRITISH PUBLIC—a first class article at a low price.

Therefore **WE HEREBY GUARANTEE** to send **CARRIAGE PAID** to anyone in Great Britain who will cut out this advertisement and send it to us together with postal orders value 35/- (thirty five shillings) one of our latest

35/- CORONATION MODEL GRAPHOPHONES COMPLETE with SIX TESTED COLUMBIA RECORDS. 35/-

The endless amusement that can be obtained from one of these charming instruments, which talk, sing, play, and, in fact, **REPRODUCE ANY SOUND THAT IS KNOWN TO MANKIND**, is coupled with the opportunity they afford for everybody to **actually hear** the FINEST MUSIC played by the VERY FINEST BANDS.

THE CORONATION MODEL plays **SOUSA'S MARCHES** played by **SOUSA'S BAND.**
THE CORONATION MODEL sings **DAN LENO'S PATTER SONGS.**
THE CORONATION MODEL plays WALTZES, POLKAS, LANCERS, &c.
THE CORONATION MODEL plays SELECTIONS FROM THE FINEST OPERAS.
THE CORONATION MODEL plays SELECTIONS FROM "THE BELLE OF NEW YORK," "SAN TOY," &c.
THE CORONATION MODEL plays CORNET SOLOS, MANDOLIN SOLOS, BANJO SOLOS.
THE CORONATION MODEL gives a performance by the MOORE AND BURGESS MINSTRELS.
THE CORONATION MODEL sings SCOTCH SONGS, WELSH SONGS, &c., &c.

THE CORONATION MODEL has no limits. IT PLAYS EVERYTHING.

Remember the price of 35/- is ridiculously low, and includes the **WHOLE INSTRUMENT COMPLETE WITH SIX RECORDS. NO EXTRAS.**

DON'T WASTE TIME THINKING ABOUT IT. **RUSH OUT** AND GET A Postal Order and send it up to us. We will at once dispatch the CORONATION MODEL to you, and you are instantly delighted.

YOU WILL QUIVER WITH DELIGHT when you hear this instrument playing and singing. EXTRA RECORDS COST YOU ONLY 1/6 POST FREE, and you can buy ONE ONLY or ONE HUNDRED.

IF YOU HAVE ALREADY A TALKING MACHINE, send for a sample of our new EXTRA FINE, HIGH-SPEED, MOULDED RECORDS, which cost 2/- each POST FREE.

IF YOU HAVE A GRAMAPHONE, send for a sample of our new DEEP CUT, EXTRA CLEAR-TONED RECORDS, which cost only 2/- POST FREE.

WHEN YOU HAVE A CORONATION MODEL your friends will WONDER WHERE you got such a lovely instrument. You won't forget to tell them, will you? First, however, GET IT, and get it from

THE SAXON TRADING CO., 84, OXFORD STREET, LONDON, W. (Trade Supplied.)

THE ORD-HUME PICTURE LIBRARY

Top right: Emile Berliner's gramophone of 1890. Discs can be mass produced and were harder, giving louder reproduction, than cylinders. leading to their success.

Right: mail order advertising has changed little, as this 1905 advertisement shows.

Flying machines

The coachman of Sir George Cayley, the 19th century aeronautical pioneer, threatened to resign because he objected to flying in his master's primitive glider. This event took place near Scarborough, Yorkshire, in 1853 and it represented one of the first steps towards the aeroplane and passenger carrying flights. It was already possible to fly using a balloon, and on the occasion of the siege of Paris 1870–1 many passengers were carried out of the beleaguered city by balloons. But balloons could fly only where the wind took them and this was a very unreliable source of power.

During the latter half of the 19th century the internal combustion engine was invented and adapted to power airships, and then in 1903 the Wright brothers made the first successful flight in a heavier-than-air machine. Flying became a popular sport of the Edwardian era, but a few farsighted people were planning to use aircraft as a means of transportation. During 1911 mail flights were introduced experimentally in a number of places, including one between Hendon, in north west London, and Windsor. But in Germany an even more important service had commenced using the new rigid airships or Zeppelins. Between 1910 and 1914 some 35,000 passengers were carried on flights between German cities without a single mishap.

Just as aircraft and airships were emerging as a practical means of transport, World War 1 broke out and they were hastily adapted for military purposes. The performance of aircraft and their engines improved rapidly under wartime conditions. A typical flying speed in 1914 was 70 miles per hour (113 km/h), but by 1918 this speed had been doubled.

When the war ended, a number of bombers were converted to carry passengers, but their useful load or *payload* was very small and this factor resulted in high fares. The first daily air service was opened in 1919 between London and Paris, using one of these bombers carrying four intrepid passengers. Although flying cut the journey time to 2½ hours, it was expensive and the single fare was 20 guineas (about £113 [$300] at today's values).

By the mid 1920s aircraft were being designed specially to carry passengers and the first true airliners emerged. One such airliner was the three engined Armstrong-Whitworth Argosy which could carry 20 passengers at 95 mph (153 km/h). Unfortunately it had a range of only 300 miles (480 km) and

PICTUREPOINT

The airfield at Hendon, a few miles from Central London, was a popular centre for flying. This Underground railway poster of 1912 depicts the displays which were a great attraction there. Many pioneers began their flying days at Hendon.

PHOTRI

The first transatlantic crossing was made by A. C. Read in 1919 in this flying boat, the NC-4, with frequent stops for fuel.

consequently it had to make many refuelling stops on a long distance route. Even the famous Handley Page HP42 of 1931, which could carry 38 passengers, was not able to fly significantly faster or further. Journey times on a short route such as London to Paris were considerably shorter than surface travel, but on long routes the advantage was not so marked. To fly from London to Cape Town took about 11 days (assuming there were no delays) whereas the mail steamer took 17 days.

The year 1936 was a significant one for air travellers. In Britain, Imperial Airways received the new Short Empire flying boats which, with a range of 800 miles, revolutionized the airline's long distance routes. Cape Town could be reached in $6\frac{1}{2}$ days; a year later this time was reduced to $4\frac{1}{2}$ days – the Earth-shrinking process had really begun. In America, Douglas DC3s were used for the first time, and this great airliner is still in service today.

Although Alcock and Brown's Vickers Vimy bomber flew across the Atlantic in 1919, it was not until 1939 that aircraft capable of carrying a useful payload across the Atlantic were ready for service—but World War 2 intervened. Once again war accelerated improvements in aircraft performance, and in particular fighter speeds which were dramatically increased due to the introduction of the jet engine. But in America piston engined transport aircraft, such as the DC4 and the Lockheed Constellation, were being developed and these made excellent airliners after the war. Britain had no such aircraft, and the decision was taken to design airliners powered by the relatively new jet or turbo-prop engines. The de Havilland Comet and Vickers Viscount were built and eventually made their mark, cutting flight times by substantial margins.

The popularity of the turbo-prop engine for large airliners was shortlived, but jet engined airliners were produced in huge numbers, with the Boeing 707 family outstripping the others. The great boom in air travel in the 1960s led to the airlines clamouring for bigger and faster airliners. So the Boeing 747 Jumbo jet, BAC/Aerospatiale Concorde and other similar aircraft were produced—a far cry from Cayley's glider.

This is the plane which inaugurated the first London–Paris airline service on 25 August 1919—a de Havilland DH16, converted from the DH9A bomber.

The three-engined Argosy airliner, shown here, was the mainstay of Imperial Airways flights between London and European capitals in the 1920s and 1930s.

RADIO TIMES HULTON PICTURE LIBRARY

MARY EVANS PICTURE LIBRARY

Radio and television

The first radio transmissions ever made were by Heinrich Hertz in about 1886. The transmission distance was a few feet and the 'message' transmitted was simply a transfer of energy which appeared as a spark in a gap in a loop of wire. When Marconi came to the subject some eight years later at the age of 20, a number of people such as Sir Oliver Lodge and A S Popov, who the Russians call the inventor of radio, were studying the new 'Hertzian waves'.

It was Marconi's vision and enthusiasm which brought the use of radio waves out of the university laboratory and into industry. Their first main use was for communication over water, where the well established telegraph and telephone could not reach. In 1897, Marconi succeeded in sending signals to a tugboat some 18 miles (29 km) away, and within a few years it became common for ships to carry the new 'wireless' apparatus.

The earliest receivers were linked to a version of the telegraph apparatus which inked the signals on to paper. When signals were strong, and where the receiver was sensitive enough, headphones could be used so that the operator could write out the morse message more quickly. A variety of types of receiver were invented, the critical component being the detector. Marconi's first receivers used a device called a *coherer*, which consisted of iron

PHOTRI

Above: the early days of broadcasting. This view of an American transmission in progress shows the paraphernalia of a 'wireless' broadcast, including dry batteries and a single early valve.

Below: the value of radio in wartime. Listening to Churchill in a London pub.

RADIO TIMES HULTON PICTURE LIBRARY

filings which clung together under the influence of radio waves, thus changing their total resistance. The coherer had to be constantly vibrated so as to shake the filings apart for constant reception. Improved detectors soon came along, the most popular design employing a crystal of a substance such as magnetite or silicon which was probed with a thin wire called a *cat's whisker*. The *crystal set* was widely used until the introduction of the thermionic valve [vacuum tube] after World War 1, and because of its simplicity and cheapness was popular with young hobbyists until the coming of the transistor.

The first broadcasts of speech by wireless were made in Germany and America in 1906 but no-one seems to have placed great importance on these sort of experiments. Ships could communicate over large distances quite well using morse, and speech, requiring more complex apparatus, was no great advantage. It was not until the triode valve was introduced by de Forest, an American, that speech and music could be transmitted easily.

The triode can be used either to transmit, receive or amplify signals and made much greater power possible. Amateurs used telephony, rather than telegraphy, increasingly from 1919 and often would broadcast records or impromptu concerts; soon it was realized that a new entertainment industry was in the air. The first regular broadcasting stations began in 1921.

Owning a radio set during the early 1920s was quite an undertaking. The signals were so weak and the receivers so insensitive that the aerial wire had to be strung up about 30 feet (10 m) above ground, usually running down the garden like a high altitude washing line, and had to be well insulated. Even so, there was only enough power in the receiver to operate headphones. The first loudspeakers were no more than earphones with megaphone cones attached, which just about gave enough output to be heard as long as everyone in the room kept quiet. Then in the mid 1920s Rice and Kellogg in the USA invented the moving coil loudspeaker, the design which is basically unchanged today. By the early 1930s, radio sets were tolerably sensitive and gave good quality sound.

Television was a logical development from radio, but there were many practical problems concerning the exact method. John Logie Baird was the first to demonstrate a working method of turning pictures into electrical signals, using electro-mechanical methods in 1926, but his system was most impractical. It was the electronic system invented by Vladimir Zworykin in the USA and perfected by a group at EMI in Britain which eventually became accepted. Transmissions on the 405 line system, using VHF (very high frequency) signals, started in Britain in 1936.

The early home TV receivers had picture tubes just five inches (12.5 cm) in diameter, though just before World War 2 tubes nine inches (23 cm) across were introduced. They consumed great amounts of power, and research after the war concentrated on making the picture larger and the electronics more compact. During the 1950s the 21 inch (53 cm) screen was regarded as a luxury.

The invention of the transistor in 1948 was not instantly recognized as revolutionary, and it took some ten years before it use became at all widespread. Nowadays all electronic equipment is compact and the only limitation in TV design is the picture tube, which has not changed in principle since its invention. The pocket TV, with solid state screen, still lies in the future.

PICTUREPOINT

Above: the ultimate in home entertainment pre-war style. This combined radio and television with 7 inch (18 cm) screen was on the British market in 1938.

Below: the apparatus that started it all. The first public demonstrations of a television system were given in 1926 using this crude mechanical scanner.

RADIO TIMES HULTON PICTURE LIBRARY

Radar

Of all the technical advances that were considered to have been decisive in World War 2, radar is perhaps the best known. Its use played a major part in the Battle of Britain, for example. Yet the principle of radar was no secret, and many experiments had been carried out in a number of countries in the years before the war.

It had been found very early in the investigation of radio waves that they could be reflected by certain materials, just like light waves, and even in 1904 there were patents for a collision prevention device which was to work by detecting the radio echoes from an object.

The technical problems of designing such a piece of apparatus are not great, since all one needs to know in a simple system is whether or not there is another vessel within stopping distance of one's own vessel. For a practical radar system, however, one needs to know also the distance and exact direction of the other object, and this presents rather greater difficulties.

The early experiments used continuous wave transmissions—in effect, a steady tone. In about 1925, however, researchers began to use pulses instead, timing the interval between transmission and reception of the reflected pulse. To start with, the experiments were carried out to discover the height of the *ionosphere*, a layer of the upper atmosphere reflective to radio waves. The interval is a small fraction of a second, so methods of displaying the information had to be devised. In this system lay the principles of true radar.

Many governments became interested in the idea of using radio pulse ranging for military purposes during the 1930s. Different governments put the emphasis on different aspects, but it was in Britain that the need for an accurate early warning system was becoming most acutely felt. The growing military power and aggressiveness of Germany made it obvious that this would sooner or later become vital, and in 1934 a committee of scientists was set up to investigate various methods.

MARCONI

MARCONI

The Chain system of radar was begun in 1937 and eventually ringed the UK. At top is the aerial array; at right is the inside of the station.

MARCONI

Above: radar installations such as this became a common sight at airports after World War 2.

Below: the components of an advanced airborne radar system, as fitted to British Lightning fighters.

FERRANTI

One of the suggestions was from a physicist at the National Physical Laboratory, Robert Watson-Watt. He set up an experiment using an existing short wave transmitter which gave good results, and began work with a team which by 1938 had set up a chain of early warning stations along the Thames estuary, later extended to cover much of southern England. It was this network which was so valuable during the Battle of Britain.

Land based radar is not limited in size: aerials and equipment can be as large as necessary. But for the apparatus to be used aboard an aircraft, small size is important. Another requirement is that the beams should be sharply defined, so that echoes are not received from the land or sea in the same general direction. With the limitation of size, this can only be achieved by using very short wavelengths, such as microwaves.

Microwaves have wavelengths of the order of a few centimetres, with frequencies of several hundred million cycles per second. In 1939, no apparatus had been designed which could handle such high frequencies at the high powers necessary for long range work. A team began work in Britain in the autumn of 1939 to produce a suitable transmitter, using various American devices as their starting point. By early 1940, they had come up with the *cavity magnetron*, which was capable of producing about 50 kW of pulsed power—hundreds of times more than had previously been possible.

At about this time, British and American expertise was pooled, and microwave radar came into use in Allied aircraft. One joint invention which was of great use was the PPI—plan position indicator—system of display—the familiar circular screen swept by a beam. This gave a clear picture of the ground below and enabled navigators to get a much more accurate fix on their position. Aboard ships, it gave a view of a whole convoy at once, while its use against submarines proved highly successful.

Meanwhile, German radar had not advanced to such a great extent. The German system was still operating on much longer wavelengths, and it was not until a cavity magnetron was discovered in a plane wreck in 1943 that the Germans had any idea that the Allies had developed microwave radar. They set about developing their own cavity magnetrons, but these were not produced until almost the very end of the war.

Transistors

To many people, the word *semiconductor* means the same as *transistor*. Yet the transistor was not invented until 1948, and did not come into widespread use until ten years after that, while the electronic properties of semiconductors were known even before radio itself was invented.

Semiconductor materials are simply those which have electrical properties midway between those of conductors and insulators. As early as 1880 it was known that certain semiconductors could act as rectifiers—that is, they would allow current to pass one way, but not the other. This property was exploited in the early days of radio, when various crystals with thin wires touching them—'cat's whiskers'—were used in the detector stage of radio sets, called crystal sets.

The coming of the valve or vacuum tube at the end of World War 1 did away with the need for the cat's whisker, but other semiconductor materials, such as selenium, were used as power rectifiers. During World War 2 it was found that crystals were more suitable than valves for detecting radio waves of very high frequency, *microwaves*, and research on them began again in earnest.

When they were first used in crystal sets, the experiments on semiconductors were carried out more or less on a trial and error basis, and no one had any real idea how they worked. By the 1940s, however, the theory had been worked out from quantum physics, which did not exist at the turn of the century. It was possible to understand the behaviour of conductors in terms of the movement of electrons between various energy levels of atoms.

Armed with this knowledge, it seemed that it ought to be possible not just to make semiconductors which would act as *diodes*, as in rectifiers, but as *triodes* as well, in which case one could make an amplifier with them. Instead of the electrons forming the current being physically accelerated through the space inside a glass tube, then controlled by stopping some of them by putting a grid in the way, it should be possible to control the flow of electrons inside a semiconductor by the effects of other electrons, a much more efficient process.

A team at the Bell Telephone Laboratories in New Jersey, headed by physicist

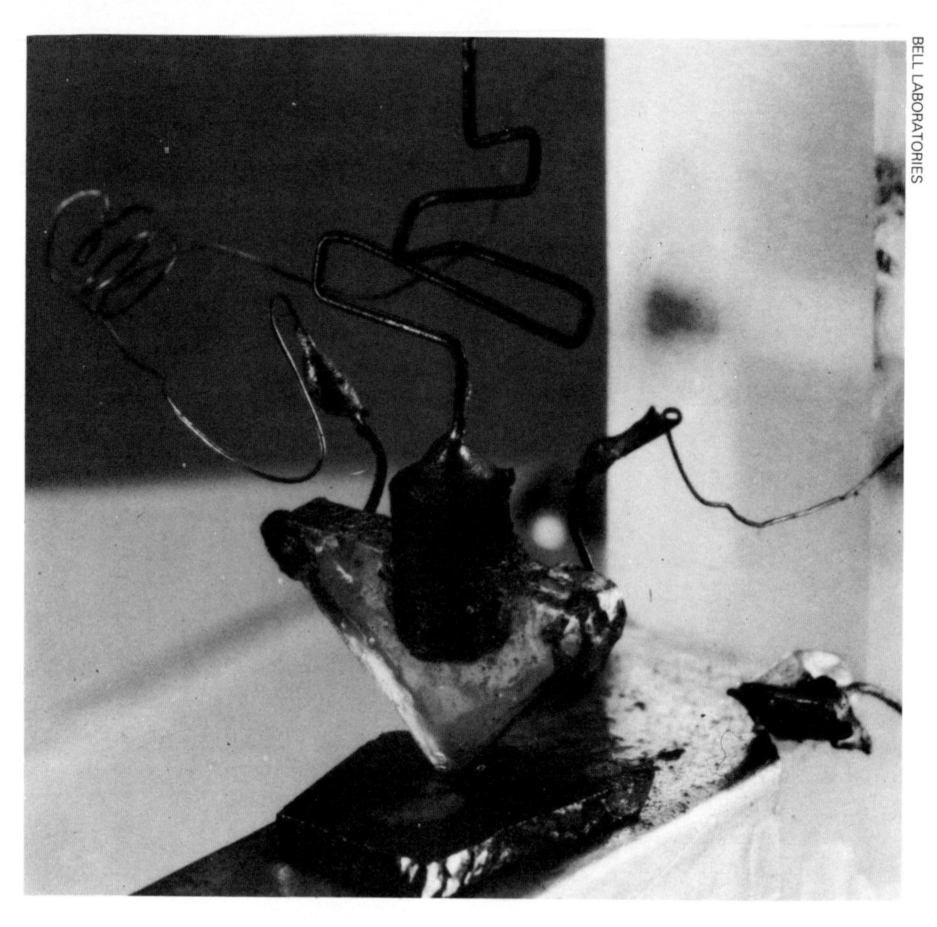

BELL LABORATORIES

Above: the first transistor ever made, a point contact device. It was imperfect and noisy, but it represented a great advance in electronics.

Below: the first junction transistor, the forerunner of today's types. The most expensive piece of equipment used in its development was an oscilloscope.

BELL LABORATORIES

William Shockley, began work with this aim in mind. To start with, they used germanium, a semiconductor material, as one plate of a capacitor hoping that the electric field of the other plate would influence the flow of electrons in the germanium. It did, but not by as much as anticipated. It seemed that the surface of the semiconductor was having some effect on the flow, and a theorist, John Bardeen, worked out a theory to account for the observed effects.

In an experiment by Walter Brattain, another of the team, it was found that improved results were obtained if the contact used was actually on the surface of the semiconductor. This knowledge, together with the discovery that semiconductors with excesses of either electrons or 'holes'—spaces for electrons—could be made by adding impurities, led to the invention of the first *point contact* transistor.

In this, two pointed contacts, the *emitter* and *collector*, were attached to a block of germanium which was mounted on an electrode called the *base*. The current between the collector and base is controlled by the bias on the emitter, thus giving amplification. Because of the small area of the point contacts, the device was not very efficient but it represented a major triumph in electronics.

Shortly afterwards, Shockley put forward the idea of the *junction transistor*, in which two types of semiconductor material are joined together in a sandwich, so that the action takes place over the whole area rather than just as two contacts. This had greatly improved performance, and led to the transistors of today.

The first transistors were demonstrated by making a radio set entirely without valves capable of being heard in an auditorium. The significance of the new invention was appreciated at the time, but since new manufacturing processes had to be invented to exploit the potential of the new components, little more was heard of transistors for a number of years. Shockley, Bardeen and Brattain were awarded a joint Nobel prize in 1956 for their discovery, but it was the introduction of the first mass produced transistorized radios in about 1955 which brought the new devices to the attention of the public.

BELL LABORATORIES

Right: this cutaway of an early lab model point contact transistor shows clearly how they were constructed. Small as transistors were, an even greater advance came with the introduction of integrated circuits in the mid 60s: early models are shown below compared with a fly's eye.

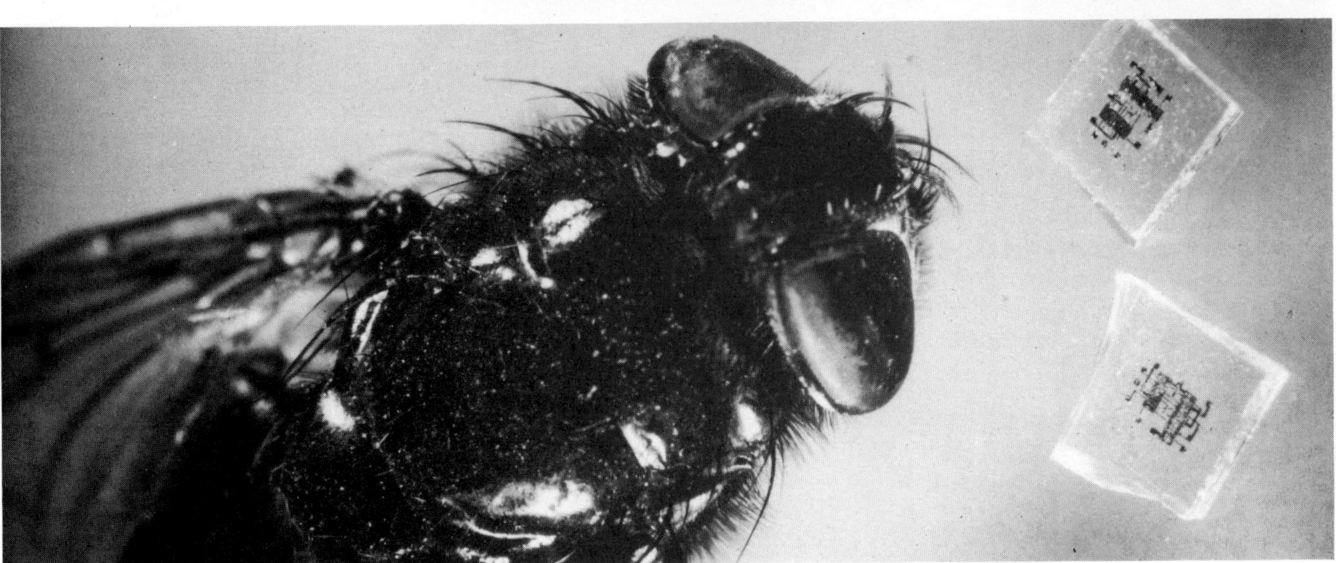

RCA

Nuclear power

Even before Einstein published his famous $E = mc^2$ equation, it had been realized that there is a link between mass and energy. Einstein's equation suggests that one can be converted into the other, and just such a phenomenon was observed around the turn of the century in terms of radioactive decay. When a radioactive element decays into another of lower atomic weight there is a release of energy. Indeed, this process is used today as a means of generating heat aboard satellites and remote transmitters. The problem was, how can the vast amounts of energy stored up in atoms be released at all, other than at the slow natural decay rate?

At the same time, astronomers trying to explain the vast energy production of stars realized that such a process must be going on, though for many years it was impossible to discover the exact reactions taking place. Until 1939, however, the notion of converting mass into energy remained a matter for speculation: no one could say whether it could ever be achieved.

The discovery which changed the situation was the identification of nuclear *fission* in Germany in late 1938. Otto Hahn and Fritz Strassman had bombarded uranium with neutrons, and obtained two products, one of which seemed to have the properties of barium—very much lighter than uranium. They could not believe this: they were expecting to obtain products almost as heavy as uranium itself. It was Lise Meitner and her nephew Otto Frisch, whom Hahn told of his results, who suggested the true explanation of what had happened. A neutron had been added to the uranium atom, which had split into two much lighter elements. The mass of the products was slightly less than that of the original, so the difference was released as energy.

The importance of the fission process is that although just one neutron begins it, two or three neutrons, on average, result. These neutrons in turn may spark off further fissions in what is called a *chain reaction*, so that the process becomes self sustaining.

If this chain reaction takes place slowly, the result is a release of power which can be used to generate electricity. If it takes place very rapidly, the result is an atomic bomb.

This potential was soon realized, and as a result of rapid communication between the scientists concerned, full

Below left: the second nuclear pile, rebuilt at Argonne using materials from the first, could give 100 kW of power.

Below: Enrico Fermi, who with Leo Szilard, a Hungarian, built the first reactor.

UKAEA

details were published in the United States within weeks of the experimental work. To start with, no one was sure whether a self-sustaining fission reaction was possible: the number of neutrons liberated by each fission was then unknown. But the likelihood was sufficiently strong for a group of American physicists to persuade Albert Einstein to sign a letter addressed to President Roosevelt. The letter reported the new results and outlined the possibility that German scientists might be working on an atom bomb. The result was the Manhattan Project, which was to explore the power of the atom.

One aspect of this was to try to set up a slow chain reaction. An important discovery in this respect had already been made by Italian Nobel prize winner Enrico Fermi who discovered that by slowing down neutrons, using some material with atoms of low atomic weight, their chances of a reaction were increased. At the same time, the neutrons should not be absorbed. Graphite turned out to be a suitable material, and in 1942, after considerable experiment and calculation, Fermi and his colleague Leo Szilard were ready to begin construction of the world's first nuclear reactor. It was to be made of layers of graphite and uranium, piled one above the other, with strips of cadmium—a good absorber of slow neutrons—as a safety measure. The site of the reactor was to be specially prepared in the Argonne National Forest, some 20 miles (32 km) outside Chicago. Unfortunately, Fermi was ready before the site was, labour difficulties during the war having caused a delay. Consequently, it was decided to build the reactor, to be called a *pile* from its appearance and as a good way of keeping its true nature secret, in a squash court within the Chicago University campus. Fermi was sure from his calculations that there would be no danger, but even so the last hours as the pile was completed were tense.

The pile, consisting of 385 tons of graphite and 40 tons of uranium and uranium oxide, filled the high room. On Wednesday, 2 December 1942, a small group of scientists assembled to watch the final stages. The pile was complete, and all that remained was to pull out a cadmium control rod slowly, watching the increase in neutron activity. A man stood by with an axe ready to cut a rope holding an emergency rod, and a squad of men carried buckets of cadmium solution in case that failed.

The final rod was pulled out, a short length at a time, while Fermi calculated the level that the counters should reach. Finally, just after 3.20 that afternoon, the counters continued to rise after the rod had been pulled out, showing that a self sustaining reaction was taking place. The total power output was a mere 0.5 watt—about the same as a small flashlight—but it represented one of the most significant steps in the history of invention.

The first full scale nuclear power station was Calder Hall, Cumbria, shown here, which opened in 1956. By 1975, over 12% of Britain's electricity was supplied by nuclear power.

UKAEA

Index

JOHN BISHOP

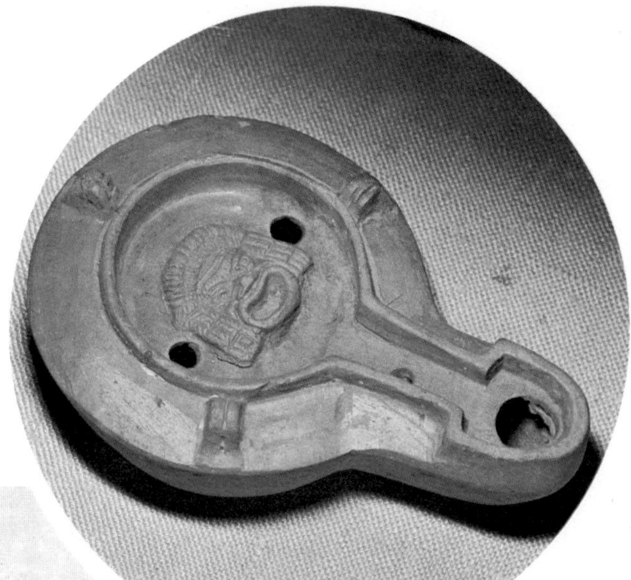
C M DIXON

RADIO TIMES HULTON PICTURE LIBRARY